An introduction to the historiography of science

AN INTRODUCTION TO

the historiography
of science

HELGE KRAGH

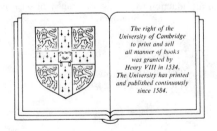

The right of the
University of Cambridge
to print and sell
all manner of books
was granted by
Henry VIII in 1534.
The University has printed
and published continuously
since 1584.

CAMBRIDGE UNIVERSITY PRESS

Cambridge

New York   Port Chester   Melbourne   Sydney

Published by the Press Syndicate of the University of Cambridge
The Pitt Building, Trumpington Street, Cambridge CB2 1RP
40 West 20th Street, New York, NY 10011, USA
10 Stamford Road, Oakleigh, Melbourne 3166, Australia

First published 1987
First paperback edition 1989

Printed in Canada

British Library cataloguing in publication data
Kragh, Helge
An Introduction to the historiography of science.
1. Science – Historiography
I. Title
507'.2   Q125

Library of Congress cataloging in publication data
Kragh, Helge, 1944–
An introduction to the historiography of science.
Bibliography
Includes index.
1. Science – Historiography.   I. Title.
Q125.K686   1987   507'.22   86-28334

ISBN 0-521-33360-1 hard covers
ISBN 0-521-38921-6 paperback

Contents

Preface

....

The subject of the present work is what I consider to be the essentials of the historiography of science. I discuss a number of problems which, I suggest, are of fundamental importance to almost any serious historical study of science, irrespective of its particular field and period. There are, of course, historiographical issues which are peculiar to certain approaches, disciplines and periods. Most of these I have left untreated or only touched lightly. Thus science before 1500 only figures sporadically in the book, and issues peculiar to the social and institutional history of science have only received scant attention. Apart from these limitations there are other important topics which I do not discuss because they are only indirectly related to the main themes of the book. These include various philosophically based views concerning the historical development of science, such as the historiographical theories of Kuhn, Lakatos and others, and also the question of the so-called driving forces of scientific development.

The structure of the book is as follows. Chapter 1 gives an outline, separated from the rest of the work, of the prehistory of history of science. The chapters 2 to 7 deal with matters of a general historiographical nature, being an introduction to theory of history as applied to history of science. As a historical discipline, history of science is amenable to the same theoretical reflections which are valid in general history. Practitioners of the discipline, whether trained as scientists or historians, should be familiar with these reflections. In chapters 8 to 10 I discuss some of the basic problems in the general historiography of science. These include problems of periodization, ideological functions and the tension between diachronical and anachronical historiography. The rest of the work deals with the critical use and analysis of history of science sources and related issues. While the analysis of sources is essentially the same for any historical discipline, in some respects the historian of science faces problems that are peculiar to his field. One such problem is the possibility of experimental reconstruction of history. The two last chapters give a critical review of versions of quantitative history of science.

An earlier Danish version was translated into English by Jean Lundskjær–Nielsen. The work has received support from the Danish Research Council for the Humanities. I gratefully acknowledge this support. The book has benefited from various suggestions and critical remarks made by two referees unknown to me.

Helge Kragh
June 1986

1

Aspects of the development of the
history of science

Although the history of science as an autonomous academic discipline only developed in the 20th century, activities that might justifiably be described as early forms of history of science have been taking place for centuries. Historical descriptions and analyses have always followed the development of science. Indeed, even a superficial consideration of the history of science in former times reveals that many of the central historiographical problems discussed in modern history of science can also be encountered in earlier centuries.

Throughout most of the period in which science developed, it was learnt and cultivated as part of a historical tradition that was indistinguishable from science proper. In Classical times and in the Middle Ages in particular, the usual form of cultivation of science involved relating to earlier thinkers. Critical commentaries and analyses of the Classical works were made and these were used as a point of departure for new thought and contributions of current interest. When Aristotle wished to say something about atoms and the void, he reproduced parts of the history of atomism and embarked on a discussion with the long-departed Democritus. When a Greek mathematician wanted to solve a problem, the natural way to proceed was to begin by giving an account of the history of that particular subject, which was regarded as an integral part of the problem.

Classical historians were interested first and foremost in contemporary history and did not consider it of much value to consider earlier events or developments in a historical perspective. This topical, and therefore in one sense, ahistorical attitude was based on the Greeks' perception of critical historical method: the only reliable sources were believed to be eye-witnesses, people who had personally been present at the event under discussion and as such could be cross-questioned about the event by the historian. As a result of this approach, Greek historical perspective was, for the main part, limited to a single generation.

Another factor that contributed to the absence of a real historical perspective was the prevailing view of time and the uncertain chronology. It was usual for the Greeks to regard time as cyclic or, as far as short periods of time were concerned, as static. This notion of time does not support the fundamental idea of historical development, according to which modern ideas and events are seen as the results of the dynamics of the past. The Greeks had no tradition for, or interest in, dating events and often made do with dating them as having happened 'long ago'. Precise dating and the placement of events in chronological order are largely bound up with a linear concept of time. A linear and dynamic view of time derives especially from Judeo–Christian thought and did not become widespread until the Middle Ages in Europe.

Our knowledge of the Classical form of history of science is greatly limited by the almost total absence of original source material. Thus, we know that Eudemus, who lived in the 4th century BC, wrote both a history of astronomy and a history of mathematics, but these works have disappeared. The knowlege that we do have comes mainly from later commentators working at the end of the Classical period or at the beginning of the Middle Ages. One example of these is Proclus (*c.* 420–485) who wrote a historical account of Euclid's mathematics. Simplicius (*c.* 540), who wrote detailed commentaries of Aristotle's works on natural philosophy and, in connection with these, also gave an account of the ideas held by earlier natural philosophers, is another example. The commentaries written by Proclus, Simplicius and others can reasonably be regarded as late-Classical history of science.

In the 16th and 17th centuries, when the new science came into being, history was still regarded as an integral part of scientific knowledge. History, especially Classical history, was regarded by pioneers from Copernicus to Harvey as definitely present and relevant to the current progress of science. During the scientific revolution the Classical authorities were often used as opponents in ideological arguments. At the same time, history served as legitimation for the new science. By referring to the great philosophers in the past, a tinge of respectability could be lent to science.

From the end of the 17th century the attitude towards the Classical authorities changed. It became common to highlight the modern world's knowledge at the expense of that of antiquity. Many of the pioneers of the new science were strongly influenced by protestant religious views: they critisized Classical Greek scholarship for being heathen, and wanted to trace science back to a Biblical knowledge dating from before the time of the Greeks. Wherever such knowledge was not known, it was constructed from the Bible. Sennert, Boyle and Newton were among the many who thought that Moses had possessed a divine insight into the laws of nature.[1] Atomism, in their view, did not owe its existence to the heathen and atheist Democritus, but to the prophet Moses. This view helped to invest atomism with social authority in the 17th century. Gradually, as science became authorized as worthy in its own right, age became unnecessary as a means of legitimation and references to the great ancestors seemed superfluous.

The historical form that decked much of earlier science is well illustrated by Joseph Priestley's *The History and Present State of Electricity* (1767) and *History and Present State of Discoveries Relating to Vision, Light and Colours* (1772). These were pioneering works of what was then front research, but they were nevertheless presented as 'histories'. Priestley was one of the many who regarded the historical development as a natural part of their science, a stocktaking of what had been achieved and of the problems that were still unresolved. In this way history was given a role in the sciences of the day. In full agreement with Priestley, the French astronomer and historian of astronomy Jean–Sylvain Bailly regarded the history of science as a report on 'what we have done and what we can do.'[2]

For Priestley and his contemporaries the history of science was primarily a tool, the value of which was bound up with the progress of the research being carried out at that time.[3]

> Great conquerors, we read, have been both animated, and also, in a great measure, formed by reading the exploits of former conquerors. Why not may the same effect be expected from the history of philosophy to philosophers? May not even more be expected in this case? . . . In this case, an intimate knowledge of what has been done before us cannot but greatly facilitate our future progress, if it be not absolutely necessary to it. These histories are evidently much more necessary in an advanced state of science, than in the infancy of it. At present philosophical discoveries are so many, and

the accounts of them are so dispersed, that it is not in the power
of any man to come at the knowledge of all that has been done, as
a foundation for his own inquiries. And this circumstance appears
to me to have very much retarded the progress of discoveries.

As a natural consequence of this attitude and the period's general
belief in progress, the history of science was unequivocally depicted
as the history of progress.[4]

> I made it a rule to myself, and I think I have constantly adhered to
> it, to take no notice of the mistakes, misapprehensions, and alterca-
> tions of electricians; . . . All the disputes which have no way contri-
> buted to the discovery of truth, I would gladly consign to eternal
> oblivion. Did it depend upon me, it should never be known to
> posterity, that there had ever been any such thing as envy, jealousy,
> or cavilling among the admirers of my favourite study.

While Priestley used the history of science in the service of contemporary science, others used it as a contribution to the debate
about the correct methodology and policy of the new science. An
early, classical example of this is Thomas Sprat's *History of the
Royal Society* from 1667. The most important aim of this work
was not to give an objective, historical account of the Royal Society's foundation, but to play a polemical and political role. In
1667, the Royal Society was only five years old as an official
institution, but it had come into being as a result of the work and
visions of a series of informal groups dating from about 1640. The
methods, ideals and forms of organization to be pursued by the
new science were the subject of much discussion around 1670.
Sprat's *History* was a contribution to this debate, directed at the
future rather than the past. Since Sprat identified some sources
(Wilkins, Boyle, Bacon and others) as the Royal Society's spiritual
ancestors and ruled out the significance of others (Descartes and
Gassendi, in particular), and since Sprat's work achieved an
authoritative status, it laid down the view of science to be followed
by the Royal Society in the future. The Royal Society, and the
activities organized in connection with it, were to be based on an
empirical view of science and not on the more deductivist ideas
adopted by such continental thinkers as Descartes.

One should note that the word 'historical' in the 17th and 18th
centuries was often used in a different sense to that in which it is
used today. A 'historical phenomenon' frequently meant a concrete,
factual phenomenon and a 'history' merely an account of the factual

conditions without it being necessary for these to belong to the past. For example, Bacon's references to 'histories' that must be researched by the future science were about concrete subjects or areas of research. We have kept this meaning of the word history in the term natural history.

The truly historical perspective that the study of the past is of value in itself and therefore not in need of legitimation with regard to the present, barely existed before the 19th century. There were, admittedly, individual thinkers, in particular the Italian philosopher Giambattista Vico (1668–1744), who emphasized the value of the historical perspective. But Vico's thoughts remained isolated throughout the 18th century which, instead, was characterized by a tendency that must be described as anti-historical. The Age of Enlightenment saw history as an instrument for progress in the battle against the old feudal order. Only the recent development was worthy of interest while the past was generally regarded as irrational and inferior. Leibniz was one of the many who believed that the study of the history of science could contribute towards an increased recognition of how scientific ideas come into existence. He viewed history of science as a contribution to the formulation of the *ars inveniendi* of which he and many others dreamt:[5]

> It is of great advantage to get to know the real sources of great discoveries, in particular of those that were made not by chance but by reflection. The result of this is not only that the history of science acknowledges what each individual has contributed (i.e. the establishing of objective historical facts) and that others are thus encouraged to acquire a similar reputation (i.e. a great model serving as an incentive), but also that the *art of discovery (ars inveniendi)* expands when one finds the path of research in outstanding examples.

Although the idea of a logic of discovery was gradually discredited, the exemplary function of the history of science – that modern research can learn from the historical elucidation of the successes and failures of earlier research – remained an important theme. A century later, William Whewell dissociated himself from the idea of a logic of discovery as understood by Leibniz. But Whewell, too, regarded the study of the history of science as justified for similar reasons. In 1837 he wrote as follows:[6]

> The examination of the steps by which our ancestors acquired our intellectual estate ... may teach us how to improve and increase

our store . . . and afford us some indication of the most promising
mode of directing our future efforts to add to its extent and com-
pleteness. To deduce such lessons from the past history of human
knowledge, was the intention which originally gave rise to the present work.

The strong belief in progress and science that was a characteristic
trait of 18th century culture was also given expression in writings
on the history of science. In the last quarter of the century, many
historical works were published, including accounts of the general
development of particular sciences, historical biographies and
accounts of shorter periods of time. Bailly wrote the history of
astronomy in a series of works between 1775 and 1782, and between 1771 and 1788 Haller published a collection of so-called
'libraries' that were historical analyses of the lives and works of
earlier scientists and physicians.[7]

   History of science in the Age of Enlightenment was marked by
a naive scientific and social optimism that was not in a position
to recognize science as a proper historical phenomenon. The strong
points in that time's history of science lay in chronological details
and surveys of the subject and not in historical reflection. The
emergence of modern science was regarded as due to the inherited
thirst for knowledge of the European race, a quality that could
only find scientific expression in connection with the revolt against
what was seen as the repressive authority of the Church. Once it
had emerged, science could not be held back and would quickly
achieve perfection. Many philosophers of the Age of Enlightenment
– including notabilities such as Diderot, Turgot and Condorcet –
thought that this state of perfection had already been reached in
physics and astronomy, with only the details remaining to be filled
in. The absence of historical consciousness was also a result of the
prevailing view of cognition, in particular of the rationalist ideas
of Descartes, which were adopted in many areas by the French
philosophers. According to Cartesian epistemology, cognition was
purely reflective and rational, a universal and ahistorical abstrac-
tion. Reason itself could not be contingent on history, which
removed the basis for a proper history of ideas and science.

   The romantic current that spread in Northern European natural
philosophy at the end of the 18th century also had some influence
on historiography of science. Romanticism in general involved a
stronger sense of history than was the norm in the 18th and 19th

centuries. Among other things, history was regarded as more relativistic, that is, the particular value and innate reason of each period and culture were recognised. Romantic thinkers often had a clear understanding of what is known as *diachronic historiography*, founded on the idea that the past should be judged on its own premises. This is revealed, for example, in their sympathetic attitude to the Middle Ages and to such unorthodox forms of knowledge as astrology and alchemy. Thus, Ørsted gave an account of medieval natural philosophy that was admittedly critical, but in contrast to the attitude that prevailed in the 18th century it was characterized by a certain amount of sympathy. 'Alchemy,' says Ørsted, 'was no randomly designed, but an absolutely essential element of the prevailing physics. All natural philosophers were searching for the philosophers' stone, for no other physics existed at that time and no other physics could arise . . . .'[8]

However, leading *Naturphilosophen* taught a view of history that was based on an intuitive, speculative insight into the spirit of the time. This was a view that was in opposition to the critical and systematic historiography that was developed at the end of the Romantic period. Accuracy, source critical methods, and responsibility as regards historical facts, were not regarded as virtues by the Romantics. Henrich Steffens (1773–1845) thought that such strivings were destructive to history as an idea. 'There are scholars of history', he wrote, 'who feel they cannot rest until they have pursued the majestic stream of turbulent history all the way to the dirtiest puddles, and this is what they call a study of sources'.[9] A similar critique was advanced in his programmatic *Philosophical Lectures*, in which a holistic approach was recommended to both the historian and the natural scientist. He had this to say about the feeling or intuition that to the true philosopher joins the whole of nature together in time and space:[10]

> Periods of time whose way of thinking, whose external existence was quite different from our own become intelligible to us by means of this. If we give ourselves up to it, we shall be renouncing that intellectual postulate of reason: to make our own age and its way of thinking into a norm for all; it will give us the organs of the times that lie hidden in the past.

As a result of the professionalization and organization of the scientific life that became established in the 19th century, a certain amount of interest arose in the history of science. But it was an interest that was primarily directed towards technical and specialist matters. The increasingly arrogant natural sciences distanced themselves from the humanities and a corresponding schism appeared between the history of science and such fields as philosophy, history of civilization and theory of history. The feeling that philosophy can learn from the history of science while the latter has nothing to learn from philosophy became widespread. This is exemplified in Whewell, who derided the examples of traditional logic as 'so trifling as to seem a mockery of truth-seeking, and so monotonous as to seem idle variations of the same theme'.[11]

The often arrogant confidence in the methods and possibilities of science that accompanied the positivistic current in the 19th century resulted in a relatively unhistorical form of history of science. By regarding the methods of science as unequivocal and universal, the historical perspective was narrowed down and interest concentrated on contemporary science and its immediate predecessors. This was explicitly stated by Justus Liebig (1803–1873), the great chemist: 'If it is *impossible* to judge *merit* and *guilt* in the field of natural science, then it is not possible in any field, and historical research becomes an idle, empty activity.'[12]

It was usual in the 18th and 19th centuries for scientists to include in their works a 'historical introduction' in which they summarized the pre-history of the subject and placed their own work in that tradition; while, at the same time, emphasizing the originality and significance of their work. One example is Darwin's 'historical survey' which he included in later editions of *The Origin of Species*. In this survey he gave a historical account and evaluation of the concept of evolution from Lamarck up to his own contributions.[13] Historical introductions of this kind are often documents that are of interest to modern historians, but they should, of course, be read critically. They often reveal more about the author than about the history of the subject concerned.

Isaac Todhunter (1820–1884), who wrote a series of histories of the mathematical and physical disciplines, may exemplify the specialist historian of science of the 19th century.[14] By virtue of their range and wealth of details alone, these impressive works are still profitably consulted today; but their technical level renders them unreadable for non-mathematicians and they can hardly be regarded as *history* of science according to modern criteria. Todhunter's works are representative of a type of history of science that has been in existence for almost 200 years: professional scientists who write about the history of their subject with regard to its contemporary status. Most of these works largely ignored (and still ignore) the historical perspective and concentrated one-sidedly on producing an accurate specialist account. Only a few outstanding scholars have been able to combine specialist expertise with a true sense and knowledge of history. Today this happy combination scarcely exists any longer.

William Whewell (1794–1866), sometimes described as the first modern historian of science, attempted to provide a comprehensive stocktaking of the historical development of the inductive sciences.[15] To Whewell, as to his period generally, science was a purely European phenomenon owing nothing to other cultures or times. But Whewell gave no explanation as to why science should be bound up with European thought, or why it arose in the 16th and 17th centuries. His purpose was rather to develop a philosophical understanding of the sciences than to understand them in their historical context. Original historical scholarship, the study of primary sources, for example, lay outside Whewell's programme, which was based on a comprehensive but somewhat random reading of contemporary sources. Instead of merely using the history of science as a collection of examples for philosophical theses, he wished to base on or even derive from history an accurate methodology of science. He maintained that history is the only acceptable source of a philosophical knowledge of science. This view is sometimes referred to as 'historicism' as opposed to the 'logicistic' view according to which logical criteria determine the philosophy of science, while history is in principle irrelevant. Whewell's contemporary, the philosopher John Stuart Mill (1806–1873), maintained a position close to logicism.[16]

Whewell's kind of history of science is representative of the philosophically orientated history that was taken up and developed later in the century, especially by scholars inspired by positivism. Mach, Berthelot, Ostwald and Duhem were all outstanding scientists who combined specialist insight with a philosophically motivated interest in the history of science. Considering the ahistorical view of science that logical positivism later made into a virtue, the extent to which early positivism made active use of the history of science in its argumentation is remarkable. Ostwald's interest in history of science revealed itself in his publication of a series of reprints of classical contributions to physics and chemistry, the so-called *Ostwald's Classic* series.[17] This series started in 1889 and, so far, comprises more than 250 volumes of original texts in translation. Ostwald's intention in publishing these volumes was to give scientists easy access to their predecessors' original publications, so that they would not be reduced to reading extracts or secondary versions of them. Twenty years later Karl Sudhoff started publishing a corresponding series of medical classics.[18]

The integration of science, philosophy and history is even more marked in Ernst Mach (1838–1916), the Austrian physicist and philosopher. Mach was of the opinion that the historical method was the one best suited to the purpose of gaining insight into scientific method. *Die Mechanik*, possibly Mach's most important work, is characteristic of his view of the history of science.[19] Mach's aim is primarily philosophical since he engages in a dialogue with the scientists of the past, by means of which he criticizes their methods and develops his own epistemology and methodology. Mach's celebrated criticism of the concept of causality and the Newtonian view of space and time is a result of this historio–critical method. The method revealed to Mach that Newtonian mechanics, far from being absolute and complete, is 'an accident of history'. Mach described his view of the function of the history of science as follows:[20]

> We shall recognize also that not only a knowledge of the ideas that have been accepted and cultivated by subsequent teachers is necessary for the historical understanding of a science, but also that the rejected and transient thoughts of the inquirers, nay even apparently erroneous notions, may be very important and very instructive. The historical investigation of the development of a science is most needful, lest the principles treasured up in it become a system of half-understood prescripts, or worse, a system of *prejudices*. Historical investigation not only promotes the understanding of that which now is, but also brings new possibilities before us, by showing that which exists to be in great measure *conventional* and *accidental*. From the higher point of view at which different paths of thought converge we may look about us with freer vision and discover routes before unknown.

A more historically conscious historiography than found in Whewell and Mach slowly began to develop from the middle of the last century. This happened under the influence of such diverse sources as Hegel, romanticism and the new historical method as developed by the Berlin School (Leopold von Ranke, Barthold Niebuhr). Among other things, Ranke (1795–1886) emphasized the objectivity and autonomy of historical knowledge and that the past had to be understood on the basis of its own and not contemporary premises. He also laid the foundation for the systematic criticism of sources with its demands for the thorough scrutiny of sources and precise referencing. The new scientific historiography was, admittedly, aimed at the historical professions of the time – mainly political and diplomatic history – and not at science which was not regarded as a historical discipline. But the standards of the Berlin school influenced a few historians of science too.

Its influence can be traced in the historiography of chemistry, to give one example. Thus Hermann Kopp (1817–1892) criticized mere chronological historiography and its tendency to show all progress in chemistry on a linear scale pointing forwards to the present.[21] His contemporary, the French historian of chemistry Ferdinand Hoefer (1811–1878) similarly made considerable use of the critical method.[22] He based his work on the study of original texts, incorporated sources from the history of medicine, art and technology and adopted a critical attitude towards progress-fixated writing. Hoefer's use of the modern critical method was not, however, typical of the 19th century, when such a basic requirement as that of giving precise references and distinguishing between primary and secondary sources was still not recognized as a necessity. Mach's *Mechanik*, mentioned earlier, is typical in this respect. Mach based his book on a comprehensive reading of original texts, but in his many quotations he does not take the trouble to indicate where the quotations come from.

In contrast to the subject centred, analytical history of individual disciplines, stands synthetic history of science in which the emphasis is placed on the unity of science and its interplay with other parts of social and cultural life. In accordance with his positivistic programme, Auguste Comte (1798–1857) argued in favour of this kind of history of science. In 1832, albeit unsuccessfully, he called for the establishment of a Chair of history of science at the Collège de France: such a Chair, the first of its kind in the world, was eventually created in 1892 and then given to a loyal follower of Comte.[23] The father of positivism wrote:[24]

It is only now that it would be meaningful to establish such a Chair, since before the present age the different branches of natural philosophy had not yet taken on their definitive character and not yet shown their various connections . . . . At this stage of our cognition, human knowledge, as far as its positive parts are concerned, can therefore be regarded as one unit and *in consequence* its history can *subsequently* be understood. But history of science, which is impossible without this unity, endeavours to make this scientific unity more complete and more distinct.

Comte's programme for a positive history of science remained, like so many of his ideas, merely a programme. Still it is important, partly because it later inspired historians, partly because it contained new thoughts. Comte thus emphasized two fundamentally different ways of presenting and understanding science, which he called the historical and the dogmatic methods. The latter is essentially the ahistorical textbook method, according to which a scientific subject is represented logically clear and distinct from other disciplines. According to Comte this is necessary for philosophical and pedagogical reasons but does not contribute to the understanding of the true nature of science. Specialist histories of individual disciplines are just as ill-suited to this purpose, for they artificially isolate the development of the *sciences* from the development of *science*, the only real object of the historical method.[25]

The so-called *historical* mode of exposition, even if it could be rigorously followed for the details of every science in particular, would remain purely hypothetical and abstract in the most important respect, for it would consider the development of that science in isolation. Far from exhibiting the true history of the science, it would tend to convey an entirely false impression of that history. Certainly I am convinced that the history of science is of the greatest importance. I even think that one does not know a science completely as long as one does not know its history. But such a study must be considered as entirely separate from the dogmatic study of science, without which the history would be unintelligible.

Thus, the relationship between the historical and the dogmatic approaches is dialectical according to Comte: in order to understand a science one has to understand its sociology and history; but a knowledge of scientific dogmatics is essential if one is to understand the history and not allow it to degenerate into a chronological pile of dead material. The dogmatic or logical order will serve as a theoretical framework for an interpretation of history.

Comte's view of the development of science had a genuine historical perspective. Although Comte's philosophy was a cultivation of progress with positivist science as its highest aim, he did not regard alchemy, astrology, cabalism etc. as mere mistakes and obstacles on the march to scientific truth. For example, he drew attention to the fact that the 'dark' Middle Ages was a necessary stage in the cultural development of mankind and ought to be evaluated sympathetically as a period existing in its own right. This rehabilitation of medieval science should be seen against the background of the 18th century massive and successful attempt to depict the Middle Ages as a *temps ténébreux* (or, as Whewell later put it, 'a mid-day slumber'). This portrayal was typical of Voltaire and the French Encyclopaedists, who thereby emphasized the peculiarity and progressiveness of the new science.

Although Comte argued for a historical approach to science, his own contributions to history of science were superficial and of doubtful value. To Comte, too, history of science was only of interest in so far as it could be related to a general philosophical system. To him, sources and historical data played a minor role, as they did for other system philosophers of the 19th century (Spencer, Mill, Hegel, Engels and Dühring, for example).

The founders of modern socialism, Marx and Engels, were clearly aware of the unhistorical and ideologically comfortable myth of the dark Middle Ages. Because of this myth, 'a rational insight into the great historical continuity was rendered impossible and history could, at most, serve as a collection of examples and illustrations for the use of the philosophers'.[26] The rudiments of a materialist history of science to be found in the works of Marx and Engels were not developed in the 19th century, when the historians mainly ignored the interrelationship of scientific development and economic and political developments. There were admittedly a few exceptions to this, especially in writings on the history of chemistry and medicine. Worthy of mention is the Anglo–German chemist Carl Schorlemmer (1834–1892), a close friend of Marx and Engels, and a supporter of Marxist socialism. Schorlemmer used parts of Marxist theory, both historical and dialectical materialism, in a work on the history of organic chemistry.[27] This was the first work in history of science that justifiably can be called Marxist and it remained the only one for half a century.

At the end of the 19th century, there was a tendency among some scientists to one-sidedly emphasize the method of science at the expense of the methods that prevailed in the humanities, including history. Eminent scientists such as Wirchow, Haeckel and Ostwald maintained that the study of history should be radically changed and subordinated to the new science dominated culture. Without in any way being influenced by Marx they spoke disdainfully of traditional 'bourgeois history' with its focusing on kings, wars and diplomacy. They wanted to replace this kind of history with a universal history based on the progress of science. Naturally, professional historians reacted strongly against what they saw as the arrogant and aggressive claims of science. In Germany, such historians as Droysen, Dilthey and Meinecke stressed that history was a humanistic discipline, a *Geisteswissenschaft*, whose methods and objectives were incompatible with those of the natural sciences. The sharp distinction which claimed to separate the two kinds of knowledge was a contributory factor towards the fact that established historians, by and large, ignored the history of science and culture. These fields were left instead to scientists and amateur historians. The history of science was, of course, assigned a central role in the German scientists' vision of a universal history of culture. The physiologist and physicist Emil Du Bois–Reymond (1818–1896) thus concluded that 'natural science is the absolute organ of culture and the history of science the proper history of mankind'.[28]

A certain amount of history of science was written out of patriotic motives, aimed at drawing attention to the excellency of the science of the nation or arguing in favour of national priority demands. Raoul Jagnaux (1845–?), for example, presented chemistry as essentially a French science. French historians and chemists engaged in an almost religious worshipping of Lavoisier, who was not only regarded as the founder of chemistry but also as a symbol of French power.[29] Many Germans minimized the historical significance of Lavoisier and emphasized instead the role of early German chemists such as Paracelsus and Stahl. This nationalistically motivated history signified that science had become an emblem of prestige, an ideological factor of national importance. History of science also played a part in the conflict between clericalism and liberalism. In several historical works the Church was accused of being an enemy of scientific progress and then, allegedly, of human progress too. [30]

It was not until the turn of the century that the scattered activities were organized and history of science began to be established as an independent profession. The first international conference was held in Paris in 1900, to be followed later by a regular series of similar congresses. Another sign of professionalization was the establishment of national societies for the study of history of science. In Germany a *Gesellschaft für Geschichte der Medizin und der Naturwissenschaften* was founded in 1901, 23 years before the American *History of Science Society* was founded. In connection with the societies, several periodicals were started for the communication of historical research. In 1902 *Mitteilungen zur Geschichte der Medizin und der Naturwissenschaften* came into being and in 1908 Karl Sudhoff (1853–1938) founded the *Archiv für Geschichte der Medizin,* usually known as just Sudhoff's Archiv. At the same time the first Chairs in history of science were established.

The professionalization of the history of medicine took place somewhat earlier than was the case with the history of science. There were regular courses on the history of medicine at several European universities from the middle of the 19th century. From 1893 a Chair in the history of medicine at Copenhagen University was held by J.J. Petersen and in 1905 the *Institut für Geschichte der Medizin* was established in Leipzig. By and large, the historiography of medicine has developed independently of the rest of history of science. It must still be regarded today as an autonomous branch with a number of problems and interests not quite shared by other fields. [31]

Paul Tannery (1843–1904) was probably the most important individual as far as the organization of the new history of science was concerned. Tannery, if anyone, is 'the true founder of the modern history of science movement'. [32] Like Comte, Tannery regarded the history of science as an integral part of the general history of mankind, not merely as a series of subdisciplines belonging to the specialist sciences. His critical attitude towards the histories of particular sciences that up till then had constituted most of the history of science, appears in the following quotation: [33]

> The scientist in so far as he is a scientist is only drawn to the history
> of the particular science that he studies himself; he will demand

that this history be written with every possible technical detail, for it is only thus that it can supply him with materials of any possible utility. But what he will particularly require is the study of the thread of ideas and the linking together of discoveries. *His* chief object is to rediscover in its original form the expression of his predecessors' actual thoughts, in order to compare them with his own; and to unravel the methods that served in the construction of current theories, in order to discover at what point and towards what goal an effort towards innovation may be made.

This theme, the relationship between the specialist history of particular disciplines and general or synthetic history of science, continues to be one of the debating points among historians.

A leading chemist and physicist as well as a philosopher of science, Pierre Duhem (1861–1916) focused on the development of the physical sciences in the Middle Ages and the Renaissance. Duhem, a devout catholic, attempted to demonstrate in a number of important works that the so-called scientific revolution was merely a natural extension of theories and methods that had already been developed by medieval scholars.[34] 'What are generally assumed to have been intellectual revolutions,' wrote Duhem, 'have almost always merely been slow, long prepared evolutions . . . . Respect for tradition is an important precondition for scientific progress.'[35] Duhem also stressed that the theories and methods of the Middle Ages owed much to the Christian world picture. His impressive project did not achieve immediate recognition and was only taken up by other historians later.

Duhem based his critical studies on a scrutiny of original texts and set new standards for precise documentation. His theory of the continuity of science and of the crucial importance of the Christian Middle Ages has not remained uncontested; but his arguments and documentation have also played a great role in modern history of science. At about the same time as Duhem, the German Emil Wohlwill worked on the same periods and problems and drew attention to the significance of the science of the late Middle Ages and the Renaissance.[36] The works of Duhem and Wohlwill later formed the basis for a school of history of science, including A.Maier, A.C.Crombie and M.Clagett, who have focused on the predecessors of the scientific revolution.

The renewal of history of science activities around the turn of the century was indebted to new discoveries in the fields of archaeology, anthropology and philology. New discoveries of source materials extended the horizons of the history of science and revealed previously unknown scientific cultures, even older than the revered Greeks. To cite just one example, the Danish philologist J.L.Heiberg (1854–1928) discovered a manuscript in Istanbul, in 1906, that led to a completely new understanding of the methods of Archimedes, in particular, and of Greek mathematics, in general.[37] In a similar way, knowledge of Egyptian and Babylonian mathematics and astronomy owed much to the decipherings made by archaeologists and philologists towards the end of the 19th century. Early Hindu mathematical sources were already being collected and studied around the year 1800 by British scholars attached to the East India Company. Ancient Egyptian mathematics was opened up from 1858, when the Scottish Egyptologist A.Henry Rhind discovered a long papyrus strip covered with mathematical examples and rules of calculation.

Another reason for the renewal of history of science was that science was just beginning to be recognized as an important historical factor, even by professional historians. J.T.Merz (1840–1922) may be singled out as representative of these early attempts to include science as part of a more general description of culture.[38] A number of wide-ranging, ambitious histories of science were written in accordance with the ideas of Tannery, attempting to chart and describe the general development of science as a whole. These works, those of Danneman and Darmstaedter for example, are impressive monuments to the ambitious current at that time, but they have not proved to be of lasting value.[39]

Finally, around the turn of the century, the history of science became the object of increasing interest because of its educational value. Many authors and teachers advocated a historically orientated method for the study of scientific disciplines. A few even practised it. In the physical sciences, the spokesmen for this were Mach and a little later Dannemann and Grimsehl.[40] In France, Duhem advocated the historical method as 'the best way, surely even the only way, to give those studying physics a correct and clear view of the very complex and living organization of this science'.[41]

We shall conclude this outline of the development of the history of science by mentioning the Belgian-American George Sarton (1884–1956). Sarton was influenced by Comte and Tannery and he wanted to institutionalize a similar view of history of science, that is, one in which synthetic unity and a belief in progress were leading elements. Sarton wrote a number of articles in which he developed his programme of what history of science ought to be,[42] and he worked hard to organize the field as an academic discipline in accordance with these guidelines. His view was, at least by modern standards, somewhat naive and surprisingly ahistorical.[43] Some of the central points in Sarton's programme are the following:

(a) The study of the science of the past is of no value in itself, but is justified solely through its relevance for contemporary and future science. The history of science can and should give inspiration to, and act as a moral for, contemporary research. Partly for this reason, it is necessary that the historian should have a good command of the modern science whose predecessors he is studying.

(b) Science is 'systematized positive knowledge, or what has been taken as such at different ages and in different places', with the accompanying theorem that 'the acquisition and systematization of positive knowledge are the only human activities which are truly cumulative and progressive'.[44] The historian ought not, admittedly, to criticize the science of the past for not living up to our current knowledge, but he ought to evaluate earlier contributions in relation to their predecessors; and when making his evaluation, he should focus on whether the development concerned constituted a step forward. The extent to which this was the case can be determined by using modern standards of progress and rationality. It is the modern historian of science who, on the basis of these standards, determines when the science of the past was based on true scientific principles and when it was merely pseudo-science. For example, Sarton refused to consider the physiological theories of Galen because he regarded them as speculative phantasies, far from the positive knowledge that ought to be a mark of science.

(c) Even though the development of science ought, in principle, to be studied as an integral part of the social and cultural currents of the age, socio–economic conditions have, nevertheless, no deep influence on the life of science. The kind of history of science practised and advocated by Sarton is internalistic. It focuses on science as an isolated, autonomous system and on the great geniuses who are the bearers of this system.

(d) When looked at in a historical perspective, science is an unmitigated good. It is the great benefactor of mankind, truly democratic and international. The study of the history of science will not only help prevent new wars, it will also build bridges between humanistic and technico–scientific cultures.

Sarton's programme was not carried out in practice and is hardly likely to be, ever. Sarton himself wrote an enormous 4,200 page 'introduction' to the history of science up to the 14th century, but neither this work nor others in the same grandiose style have had significant importance for modern history of science.[45] In practice, historians have turned away from the ideals that Sarton stressed, ideas that are mostly heard today at congresses and other ritual occasions. Sarton's enduring contribution to the history of science was, in particular, his energetic and largely successful attempt to confer on the discipline the status of a recognized academic profession. He was a tireless propagandist for the history of science and succeeded in uniting scientists, humanists and administrators in interest for the subject. One is tempted to call him the Bacon of history of science. But not its Newton.

Sarton's most important contributions took place in the United States, where history of science had been taught at a few universities since the end of the 19th century and where the ideological climate was sympathetic to his visions. This early American interest was bound up with the desire to attract students to the progressive natural sciences. To a considerable extent it had a propagandist, missionary strain. History of science was to serve a moral purpose, to be the shining account of the triumphal progress of scientific reason throughout the whole world. A prospectus from 1914, the year before Sarton arrived in the USA, states that 'a survey of the sciences tends to increase mutual respect and to heighten humanitarian sentiment. The history of the sciences can be taught to people of all creeds and colors, and cannot fail to enchance in the breast of every young man or woman, a faith in human progress and good will to all mankind'.[46]

Naturally, Sarton was not the sole organizer of the new history of science movement. Charles Singer (1876–1960), at least, also should be mentioned. He was responsible for the establishment of a department of History and Methods of Science at University College, London, in 1923. Singer's view of the history of science largely coincided with Sarton's.

We shall conclude our outline of the history of the history of science at this point. Parts of the later development will be discussed in the chapters that follow.

2

History of science

It is customary to distinguish between two different levels or meanings of the term 'history'. History (H_1) can describe the actual phenomena or events that occurred in the past; that is, objective history. In such expressions, for example, as 'throughout history mankind's knowledge of nature has always increased', history is to be understood as 'the past' or the phenomena that actually occurred in the past. But since we only have, and only ever will have, a limited knowledge of the reality of the past, most of what actually took place in the past will forever be beyond our grasp. The part of history (H_1) that we do know is not just limited in extent but is also the product of a research process that includes the selections, interpretations and hypotheses of the historian. We do not have direct access to H_1, only to parts of H_1 that have been transmitted via various sources.

The term history (H_2) is also used of the analysis of historical actuality (H_1), that is, of historical research and its results. The object of history (H_2) is thus history (H_1) in the same way as the object of natural science is nature. Just as our (scientific) knowledge of nature is limited to the research results of science that *are* not nature but a theoretical interpretation of it, so our knowledge of the events of the past is limited to the results of history (H_2) that *are* not the past but a theoretical interpretation of it. Radically positivist philosophers have maintained that the existence of an objective nature is a meaningless fiction and that it is impossible to distinguish between nature and our knowledge of it. In the same way, some idealist historians maintained that the distinction between H_1 and H_2 is a fiction that serves no useful purpose; that there is no actual history apart from that which the historian constructs from his sources.[1] There is no need, however, in the present context, for us to take this idealist view of history seriously.

Even if one did, it would hardly make much practical difference for historical research.

The term historiography is often used of H_2, meaning writings about history. In practice, historiography can have two meanings. It can simply mean (professional) writing about history, that is, accounts of the events of the past as written by historians; but it can also mean theory or philosophy of history, that is, theoretical reflections on the nature of history (H_2). In its latter meaning, historiography is, therefore, a meta-discipline, whose object is H_2; purely descriptive history will not itself be historiography but it can be the object of historiographical analysis.

History is concerned with human activities, preferably those that are socially relevant. Non-human factors are naturally included in history in so far as they have influenced human activities. If one is interested in the history of agriculture in the late Middle Ages, for example, one has to take into consideration climatic variations during that period. Climate reveals a temporal but not a historical development. When one talks of the history of climate or the history of stars it is in a different, more trivial sense than history proper which is exclusively bound up with human behaviour and consciousness. According to Olaf Pedersen, the history of science is not especially concerned with 'historical' problems in the sense used here. 'History', he says, 'is just the study of a development in time of some human event or other through a series of successive conditions, . . . one constructs a historical viewpoint as soon as one starts to organize the events with time as the parameter.'[2] This attitude, however, does not capture the distinctive character of history and does not fully cover historical practice. A mere chronological account of the various phases of an event ('I awoke at 6.30am, ate breakfast at 7am, went to work at 7.40am, . . . ') is not history. On the other hand, historical studies might well include a dipping into the past, in which temporal organization is either not included or is of no importance.

According to many historians, phenomena should be able to be described in their individuality conditioned by time and place in order to be specifically historical. What lies behind this formula is the idea that events that are historical are unique in time and space because of their location in the past. Niels Bohr was born in Copenhagen in 1885 and this event is unique in that it cannot be repeated or generalized. The demand for individuality conditioned by time and place is, however, too weak in that it does not effectively demarcate the historical sciences in relation to the natural sciences (not to mention the social sciences). This methodological demarcation is, on the whole, difficult and cannot, at any rate, be based on the idea of historical events as being exclusively fixed in time and space. And it is too strong in that it implicitly confines history to concerning itself with particular events that can be located at a particular time in a particular place. History is also concerned with law-like phenomena, relationships, trends, analogies and structures that cannot be reduced to an aggregate of individual events and that are not fixed in time and space. Such statements as 'technological innovations create economic growth' and 'the philosophy of the 17th century was dominated by empiricist ideas' will normally be regarded as meaningful historical statements. One of the effects of, and perhaps the motivation behind, confining history to unique events is that one isolates historical science from sociological, psychological and economic viewpoints; an isolation that admittedly will confer autonomy on history, but the price of that autonomy will be sterility.

The second part of the term history of science is concerned with the particular kind of human behaviour called science. When discussing this it can, again, be useful to distinguish between two levels.[3] Science (S_1) can be regarded as a collection of empirical and formal statements about nature, the theories and data that, at a given moment in time, comprise accepted scientific knowledge. According to this view science will typically be a finished product, as it appears in textbooks and articles. Since S_1 is not really conceived as human behaviour, it is not the kind of science that would be likely to appeal to the historian.

The science (S_2) that is historically relevant consists of the activities or behaviour of the scientists, including factors of importance to this, in so far as these activities have been connected with scientific endeavours. Thus, S_2 is science as human behaviour whether or not this behaviour leads to true, objective knowledge about nature. S_2 encompasses S_1 as the result of a process but the process itself is not reflected in S_1. Usually S_2 cannot be found in articles or books, but has to be pieced together with the use of historical sources.

The distinction between S_1 and S_2 corresponds, by and large, with the question of how far the emphasis should be placed on history or science. If history of *science* is meant then the science concerned will often be science in the S_1 sense, consisting mainly of a technical analysis of the contents of scientific publications placed in a historical framework. *History* of science, however, will be science in the S_2 sense. The discussion about the two forms of history of science has sometimes been conducted as though it were a debate about the extent to which the historian of science, in order to carry out his job properly, should necessarily have a good command of the technical side of the science about which he is writing; and in particular, about how far he should have a good command of the science in question in its modern formulation.

According to Pearce Williams, the modern historian of science is primarily a historian and hence need not master *all* the technical aspects of the science he is studying. The focus should be on historical and social relations while the technical details are of minor importance.[4] This view is undoubtedly shared by many leading historians of science. But there are also those who stress that history of science cannot be cultivated as though the content of science does not matter. Some authors have nothing but scorn for the historians who, through their lack of specialist knowledge, are debarred from a complete understanding of the technical aspects. 'Most historians of science . . . seldom understand the science they write about, and so they read the prefaces to scientific works and ignore the work themselves. Unless they are mathematicians, too, they have no right to meddle in the history of mathematics and theoretical physics.'[5] Kuhn, too, has criticized the neglect of concrete technical problems by certain historians.[6] But at the same time he has stressed the sterile, anachronistic nature of much science-centred history. Kuhn and Pearce Williams are among those historians of science who have demonstrated in practice that the two aspects need not be mutually exclusive.

There are so many aspects of, and accessions to the history of science (in its HS_2 sense) that there is room for, and a need for, the whole spectrum of contributions from purely technical analyses to purely historical ones. Because science is such a complex structure, history of science will necessarily be a multifaceted subject. Let us look at a subject like 'Science and Nazism' that obviously belongs to the history of science. The German Nazism of the period 1933–1945 is not particularly reflected in the science of that period, if science is understood in its S_1 sense; but it is strongly reflected in the S_2 kind of science, whose possibilities, methods and forms of exposition it influenced considerably. It would be absurd to maintain that Nazism was irrelevant when considering German science in particular. The significance of Nazism for German science cannot be captured by means of pure history of *science* (HS$_1$) but it can be caught to a certain degree by means of *history* of science (HS$_2$) even though this might ignore the technical aspects. In fact, the pioneer work on Nazism and science was written by a historian with no scientific background whatsoever.[7] The main point is that history of science is separate from science itself. As paradoxically expressed by Canguilhem, 'the object of the history of science has nothing to do with the object of science'.[8]

The fact that Beyerchen, Butterfield and many others have succeeded in writing valuable histories of science without themselves being masters of the science about which they write cannot be generalized. In other cases the tendency to ignore the content of science will prove disastrous. When it can be done depends largely on the topic being treated and the perspective of the study. In general, the closer one gets to the scientific subject the more dangerous it will be to merely consider it from without.[9]

Whatever its focus, history of science deals with science in its historical dimension. But which occurrences can reasonably be counted as both 'scientific' and 'historical' and hence be included in the history of science?

To search for a definition of 'science' or 'scientist' is hardly profitable in a historical context. Demarcation criteria, such as those to be found in philosophy of science, are mostly based on reflections on modern physical science and would be unsuited to historical use. This would inevitably lead to distortions and anachronisms and to the exclusion of forms of science that are not accepted today.[10] The view of science that we have today is itself the product of a historical process, a struggle in which only the victorious views have survived. The historian should primarily concern himself with those occurrences that were recognized at the time as belonging to the field of science, whether or not these occurrences fit in with contemporary views. But this relativistic version of what science is seems to assume that in the past, too, something called science existed. This is an assumption that is not valid for all times and cultures. Science as an institution and profession with its own norms and values mainly stems from the last century and it is only from that time onwards that one can talk of science in the modern sense of the word.

The word *scientist*, in English, is only 150 years old. Before then the profession of scientists did not really exist, which is reflected in the variety of names given to those concerned with discovering the secrets of nature: savant, natural philosopher, man of science, virtuoso, cultivator of science, and so on. It was not until the middle of the 19th century that it was felt necessary in England, for practical reasons, to find a name for the professional man of science who emerged as a social phenomenon at that time. Whewell suggested the name *scientist*, in 1834, half jokingly, and without being taken seriously. When Whewell and some others suggested the word again around 1840, considerable opposition was aroused and it was only gradually that the word became accepted as part of general speech. *Scientist* had a low status among established scholars, especially among those from the upper class because it was associated with a modern money-for-knowledge attitude. By British gentlemen scholars it was seen as a form of treason against the ideals and social values of science. Even as late as the 1890s many men of science, including such eminent people as Huxley, Kelvin and Rayleigh, refused to use the word.[11]

Further back in time, it would be even more dangerous to talk of a scientific institution or to crystallize the term 'science' from its actual context. The 'astronomers' and 'mathematicians' who lived in ancient Babylon were only scientists if one isolates and interprets their scientific activities without reference to the institutional (social and religious) context of which they can be reconstructed as elements. They did not regard themselves as scientists, still less as astronomers and mathematicians. Even so, historians of science, for the sake of simplicity and for want of better expressions, will often describe them as scientists.

The agents of the history of science are the individuals who have, in fact, helped to collect knowledge about nature or what has been thought to be so. Not all of them are scientists, a term that ought primarily to be reserved for 'individuals who did an historically appreciable quantity of original research into natural phenomena and for whom such research was an important component of their historical identity'.[12] The individuals who are relevant to the history of science include professional scientists, amateur scientists, philosophers, theologians, artisans and many others. It

is obvious that not all who have contributed to our knowledge of
nature through the ages are also of interest from the point of view
of history of science. Historians select only a small number of the
individuals who are potentially historical, to make them properly
historical. Due to the complexity of science and its history, it is
not possible to demarcate in abstract those individuals who belong
to the history of science. Nevertheless, the question is of some
practical relevance in connection with dictionaries, for example.
The authoritative multivolume *Dictionary of Scientific Biography*
thus includes 'those figures whose contributions to science were
sufficiently distinctive to make an identifiable difference to the
profession or community of knowledge'.[13] These include scientists
as well as non-scientists.

The problems of demarcation are relevant for those activities
and methods that either clash strongly with contemporary science
or border on them. Technology should be mentioned in connection
with the latter group. Although science and technology are indeed
different areas, there is not, and ought not to be, a sharp distinction
between the history of science and the history of technology. It
would be unhistorical to divide Leonardo, Smeaton, Watt or Per-
kins into (at least) two people each, a technologist and a scientist,
and to treat them as separate individuals. All the more so since
the distinction between science and technology is a relatively new
one. It is just that explicitly technological innovations do not belong
to the proper domain of history of science. The history of technology is too important to be treated as an appendix of the history
of science. It ought, first of all, to be treated as an independent
subject, worthy of study in its own right. Happily, there has recently
been an increase of interest in doing so.[14]

When evaluating the first mentioned group of activities, which
might typically include occult, religious and pseudo-scientific areas,
one must equally accept these as belonging to the history of science
to the extent in which, wittingly or unwittingly, they have contributed to the development of science. There has recently been a
clear tendency to include non-scientific activities in the history of
science, although there is some disagreement as to how far this
ought to happen. I shall illustrate the problem with an example
from research on Newton, one of the classic foci of the discipline.

Newton, if anyone the personification of science, used a considerable amount of his resources working with subjects that are

decidedly unscientific: chronology of the Scriptures, alchemy, occult medicine and prophecies of history. Manuscripts and other sources show that Newton must have used more time on these dubious works than on the mathematical and physical works on which his fame rests. One might now ask whether Newton's works on alchemy, for example, form a legitimate part of the history of science.

Newtonian research has traditionally sought to give a glorified, rationalist picture of Newton and has rather one-sidedly focused on his purely mathematical and physical works. Although Newton's (unpublished) works on alchemy have been known for a long time, scholars were disinclined to pay serious attention to them as part of the Newton that was of interest to the history of science. The evidence was either suppressed, rationalized as chemistry or explained away as a harmless hobby.[15] Since the discovery of new sources and the emergence of a reinforced research on Newton it has become impossible to deny that Newton worked long and seriously on problems of alchemy. Newton did not transcribe alchemical works merely for the sake of extracting their rational, chemical nucleus; his interest was not just a youthful fad that disappeared as he got older, and it was not the result of senility.[16] There have been three main types of answer given to the question of how far Newton's alchemy should be taken seriously as a suitable subject for study in the history of science.

Some eminent Newton scholars who represent the rationalist and science-centred approach to history of science, have denied that Newton was an alchemist at all in the proper sense of the word.[17] They have stressed the fact that his engagement was a 'private' matter and had no connection with his great scientific works. Since these are the works that are central to Newtonian research in so far as it belongs to history of science, Newton's interest in alchemy need not unduly concern the historian of science. Consequently, such recognized historians as M.Boas Hall, A.Rupert Hall, I.B.Cohen and D.T.Whiteside consider it justifiable to 'de-alchemize' Newton.

Other experts argue that Newton was indeed an alchemist, in any reasonable interpretation of the word and that he was greatly influenced by Neo-Platonic and hermetic currents of that time.[18] These scholars (P.M.Rattansi, R.Westfall, B.Dobbs, F.E.Manuel among others) think that alchemy was an integral part of Newton's world picture and, as such, consistent with the philosophy on which his works in physics builded. Newton's alchemy belongs to the history of science in its own right; not primarily because alchemy can help to throw light on certain passages in *Principia* or *Opticks*, Newton's main works in physics, but, because it was an important element in cultural history, to which Newton, too, contributed in an interesting way.

The interest in Newtonian alchemy can also be justified by arguing that it was of direct relevance to Newton's scientific theories. According to Karin Figala, Newton's alchemy is really a rational theory of matter decked out in the symbolic language of the occult sciences; at the same time a rough draft of and a further development of the published thoughts of the structure of matter.[19] As such, Newton's alchemy assumes the form of a rational, scientific theory and becomes a natural element of the history of science.

No matter how one ought to interpret Newton's alchemical works, it would be wrong to ignore them without close analysis. 'If we are going to study the manuscripts, we have to study them all, and accept what is in them whether or not it accords with twentieth century views. To say that Newton was a practising alchemist, one has neither to be an occultist himself nor to deny the abiding reality of the *Principia*. One has only to accept the manifest import of manuscripts as authentic as the mathematical papers and more extensive.'[20]

The unreasonableness of a stringent separation of an individual's scientific and non-scientific activities does not merely arise from the problems it creates concerning explanations of the origins of scientific ideas. It often also creates problems concerning the understanding of the substance of the ideas, their cultural context and content. To the English men of science in the 17th century, religious, moral and political considerations not only played a part as inspiration but also as justification. Boyle and his circle regarded the explanation of the pneumatic experiments of the day (Torricelli's, for example) as being of outright moral significance and adapted their evaluations accordingly.[21] In such a case it will be highly misleading to isolate scientific from non-scientific components. When there is documentary evidence that Boyle regarded his science as an element in the cultural struggle of his time, we cannot neglect this aspect by pleading that the behaviour of gases under low pressure cannot possibly have anything to do with the moral condition of society.

As far as the temporal demarcation of the history of science is concerned, this is a problem that history of science shares with history in general. It is mainly a question of how far there are any upper or lower temporal boundaries for history. Historians have traditionally drawn a line between so-called historical and prehistorical times, the difference being that written sources are unknown for prehistorical times. But there is agreement among historians today that this line is of no great significance and that it breaks historical continuity in an artificial way. Megalithic monuments such as Stonehenge, for example, have probably been used for astronomical purposes. In so far as this is correct the monuments are evidence of early scientific activity. The oldest part of Stonehenge dates from 2700 BC and thus, many scientists believe, forms part of the history of science.[22] The date at which one will allow history of science to begin depends on what sources are available and on how flexibly one wishes to interpret the term science. Gordon Childe is willing to attribute scientific activities to people who lived before *Homo sapiens* on the grounds that the manufacture of tools is an embryonic form of science. 'It may seem an exaggeration, but it is yet true to say that any tool is an embodiment of science. For it is a practical application of remembered, compared, and collected experiences of the same kind as are systematized and summarized in scientific formulas, descriptions and prescriptions.'[23] Whether one accepts Stonehenge or Neolithic knowledge of nature as belonging to the history of science or not, is of no particular importance. It does not really matter whether such phenomena are studied by historians of science, archaeologists or ethnologists as long as they are studied.

There is no natural upper time limit for the history of science. Although, traditionally, history deals with the past, it is hard to find convincing arguments why the present should not be amenable to historical treatment. In fact, in recent years there has been an increasing tendency towards writing historically even about current or very recent scientific activities.

It is sometimes argued that contemporary history of science is an illegitimate term. The following are some of the common objections: (1) Contemporary history (of science) concerns living scientists and their results, and draws mainly on the recollections and written statements of living scientists. The historian of our own times who relies on these sources will find it difficult to achieve a sufficiently objective distance to his material, his analyses will be 'coloured' and marked by the personal commitment the scientist has to his work. According to Collingwood, history is only bound up with such activities that *cannot* be remembered. ' . . . the past only requires historical investigation so far as it is not and cannot be remembered. If it could be remembered, there would be no need of historians.'[24] Collingwood's view of history thus rules out contemporary history. (2) In the case of controversial contemporary activities, such as priority conflicts or politically controversial science, the commitment and personal situation of the historian will influence his writings. (3) In contemporary history many of the sequences of events being studied will not yet have finished so that the historian does not know the result and is therefore unable to use it in his evaluation of the events.

None of these objections, however, are acceptable. That the source materials are contemporary do not make them less reliable or more difficult to appraise critically than is the case with many older sources. The absence of innate objectivity in source materials is not confined to the present when, on the contrary, the historian has further possibilities for checking the reliability of his sources (see chapter 13). The historian's subjective commitment is always present in good history, even when it is about earlier periods. The historian of science working on the role of the catholic church in the development of Copernicus's theories can be just as committed as one who is working on the role of American chemists during the Vietnam War. The demand that the historian's own opinions ought not to influence his work is, in any case, a misapprehension. So we must clearly dissociate ourselves from the view that modern science cannot be historically analysed as expressed by Forbes and Dijksterhuis:[25]

> The historical method is different from the systematic method. Above all it demands the ability to view with detachment the events one has to deal with. . . . This means, for one thing, that the whole of what is known as modern science, which may be defined as everything that has occurred since 1900, has had to be excluded.

As far as the third objection is concerned, it builds on the false assumption that the historian must, in a manner of speaking, be in possession of some kind of answer sheet of those events capable of being analysed historically.[26] Although it is not the task of the historian to evaluate events in relation to what today is recognized as true or false, the objection may be relevant in connection with the use of certain historiographical and philosophical frameworks. For example, some historiographical theories rely on concepts (such as crisis, success, revolution, progress and degeneration) that only make sense over a longer period of time. These schemes, proposed by Kuhn, Lakatos and others, are not immediately applicable to the most recent science.[27]

The objections against history of contemporary science are sometimes connected with the claim that no special historical insight or techniques are necessary for the understanding of the dynamics of modern science. This view has been put forward by Ronald Giere:[28]

> . . . it does not follow that history of science, as history, is crucial, except in cases where the theory in question is one held in the past. Suppose, for example, that properly to assess the evidence in 1953 for the existence and character of DNA one had to look at the development of that theory from 1945 to 1953. This would not require the special talents of a historian of science . . . surely the study of recent developments in science requires no peculiarly historical techniques – or at least not the techniques now taught by some historians of science.

However, the only way in which one can gain proper insight in the actual dynamics of modern science is by means of historical analysis; an analysis that will not only be historical in the sense that it considers a science in its time dimension, but also in the sense that it uses the techniques and methods that characterize historical research. In practice, the considerable body of literature on the history of contemporary science refutes Giere's assertion.

3

Objectives and justification

The development of history of science during the last three decades has been characterized by a proliferation of methods and perspectives rather than by the emergence of a consensus as to what, exactly, constitutes the discipline. The eclecticism and the fact that the discipline includes separate, partly conflicting, interests makes it problematic to talk about the aim of *the* history of science. Nonetheless, many have taken on the task of stating what the superior aim of the discipline should be. In what follows we discuss some frequently articulated viewpoints. In chapter 10 we shall discuss the ideological role that history of science may play in connection with scientific disciplines and institutions.

I. It is sometimes asserted that history of science, when properly conducted, can have a beneficial influence on the science of today. In its most primitive form it is suggested that the practising scientist may profitably make direct use of the history of his science; that by studying the works of earlier scientists he may receive inspiration for a solution he is looking for or even find out that the solution had already been discovered by a predecessor in the discipline. Opinions of this kind were common in early history of science (cf. chapter 1), although it was difficult to find concrete examples of scientists who had directly benefited from their knowledge of history. Truesdell is one of the few modern historians who has dared to make the same assertion. 'Knowledge of the history of mechanics can lead to new discoveries in the mechanics of today', Truesdell says, though admitting that 'no one could justly contend that even a faint idea of the true historical development of mechanics is *necessary* in order to do good research today.'[1]

A slightly different version of the thesis is the assertion that history of science ought to function as an analytical instrument for the critical evaluation of methods and concepts that appear in modern science. As mentioned in chapter 1, this was a favourite idea in Mach. More or less explicitly, the idea lies behind the many works that give a critical historical analysis of central concepts (such as space, time, evolution or causality) as they have appeared through the ages. In the opinion of Max Jammer, the works of the past cannot directly convey the insight that the modern scientist requires. It is the task of the competent historian to analyse the problems of the past in such a way that they become accessible to, and relevant for, the modern scientist.[2]

> The historical research is not regarded as an end in itself . . . . A critical historical analysis of the classical concepts and definitions of mass . . . will lead, it is hoped, to a profounder comprehension of the meaning of the term and to a higher level of understanding of its role and significance in physics.

The view that fundamental scientific concepts can only be correctly understood via the critical–historical method is widespread. For example, it was adopted by the great physicist Erwin Schrödinger (1887–1961) who made a thorough study of Greek natural philosophy in order to clarify conceptual problems of modern physics.[3] But, again, it is difficult to point out concrete examples in which the mechanism suggested by Jammer has succeeded.

Even if scientists in some cases have been inspired by historical reading, this cannot be regarded as support for the thesis of direct scientific relevance of the history of science. It is rather a sporadic influence, like the one occasionally provided by, for example, literature or religion. Even in the very few cases where the reading of the 19th century mathematician Cauchy, for example, has proved to be a promotor of new scientific insight (Truesdell's example), it is not really history of science that has had a transmitting effect. The works of Cauchy are not, in themselves, history of science. Their possible importance for modern research is not due to the historian of science but to Cauchy.

II. According to Hooykaas, history of science has at least three separate aims:[4]

> History of science provides material for a critical self-examination of science: it increases the appreciation of what we possess now, when we recognize the difficulties it cost to acquire it. It bridges the gap between science and the humanities, demonstrating how natural sciences are part of the humanism of our age. There will always be scientists who are not satisfied with knowing the contents

of theories, but who want to know their genesis and who will find this an intellectual and aesthetic pleasure.

The introductory sentence of this quotation states that because of the history of science we value our modern science more so that its social value and prestige are raised. Behind this idea of history of science as a contributor to the social prestige of science lies the assumption that this prestige is not automatically secured and may, therefore, be in need of support. The idea is further developed in James Conant to whom history of science serves as an argument for more science (that is, more money for science). Conant was one of the leading figures in American university life and research policy in the years that followed the Second World War, when history of science had just begun to be professionalized. The increasing interest in the history of science in the USA was caused to a great extent by the general degree of interest that the science-based technology of the war attracted to itself. By means of a series of historical case-studies Conant argued that the study of the science of the past leads to the conclusion that ' . . . a nation, in order to lead in technology and thus provide for the welfare and safety of its people, must lead in pure science. Thus in a few words one may sum up a long story which provides the compelling answer to the question "why more science?"'.[5]

Justifications of the type given by Conant are not at all unusual among those historians of science who regard their discipline as an integral part of a larger project, whose purpose is to understand and apply science in present and future contexts. They are especially popular with Eastern European researchers. In an article from 1975, the eminent Soviet historian of science Mikulinsky formulates this aim using expressions that Sarton would have loved to hear:[6]

> . . . the reconstruction of the past ceases to be the ultimate end of the historical research and becomes one of the stages on the way to its achievement. The main goal of research is now the understanding of regularities of the development of science, conditions and factors favouring it, for nothing can be more useful in tracing the movement from the past through the present and to the future than the knowledge of the regularities of the development of the object.

In the kinds of justification that Conant and Mikulinsky, each in his own way, represent, it is not really history of science as such whose case is being argued, but science, especially pure science.

The view is based on the fact that there is a close causal relation between pure science and technology and that history justifies such a relationship. In both Mikulinsky's and Conant's arguments it is taken for granted that science-based technology is a social good.[7]

History of science can be used, and ought to be used, to analyse the interplay between science, technology and society. But empirical evidence that science always or just usually results in technology is weak. By making use of examples from history one could easily construct a case for arguing that science does not, as a rule, lead to technology or that science and technology do not normally contribute to people's welfare and safety. In any case, no matter what the relationships between science, technology and society are and have been, history of science should not be misused as propaganda for the gospel of the scientific society.

III. History of science serves an important function as background for other meta-scientific studies, such as philosophy and sociology of science. As far as the functions that history of science can have for philosophy are concerned, there are, roughly speaking, two types. The philosopher can use history of science inductively, so that, from his knowledge of how eminent scientists have thought and acted, he can generalize these historical experiences in philosophical doctrines. This was Whewell's programme. Or, conversely, philosophical doctrines can be tested by comparing them with history of science data. History of science comes to act as a source of inspiration or as an instrument of control. In recent years the links between philosophy of science and history of science have become steadily stronger and there is no doubt that history does, in fact, play an important philosophical role. Still, the relationship between history of science and philosophy is complex and far from fully agreed upon.[8]

The relevance of history of science to such related fields as sociology of science and theory of science is analogous, in many ways, with its relationship to philosophy. Again, there is no doubt that history of science plays an increasingly important part in these disciplines. Some researchers prefer to regard history of science as one element in an interdisciplinary research programme, *science of science*, that includes all studies that have to do with science. Such a view as this will have consequences for the programme of history of science, which will be pivoted away from antiquarian or academic interests towards a more pragmatic, activist orientation. According to Günter Kröber, an East German spokesman for science of science, this is an advantage for the history of science. 'This correlation between history of science and science of science does not result in the former becoming superficially pragmatic; on the contrary, it helps to raise its theoretical level.'[9] But Kröber, too, admits that history of science is other and more than merely an element of science of science: 'It would be completely erroneous to talk of a simple incorporation of history of science writing under science of science or to imagine that it should be relegated to being a servant of science of science.'[10]

IV. As already argued by Duhem, history of science may serve an important didactic function in demonstrating the true nature of scientific knowledge (cf. chapter 1). There are many arguments, good and bad, used in favour of a historically orientated science teaching.[11] Some of these are propagandist in the sense that the use of history is justified through its alleged ability to present the sciences in a 'softer' way and thereby make them more attractive at a time when they are regarded with suspicion by many young people. One author reasons in favour of the historical method as follows: 'Many imaginative and intelligent pupils must in the past have been repelled from science, and especially from physics, because the material presented to them seemed to be cut and dried and devoid of human interest.'[12]

History of science can undoubtedly play a positive part in teaching. It can contribute towards a less dogmatic conception of science and scientific methods and it can act as an antidote to orthodoxy and uncritical enthusiasm for science. But not all teaching on history of science will play this part and certainly not automatically. History of science can be used just as well to support dogmas and to strengthen scientific authority. In general, the question of the didactic importance of history of science is problematic and the value of the historical method seems often to have been exaggerated.[13]

V. To Sarton, history of science was to reflect the humanist placement of science, the 'centre of human evolution and its highest goal'.[14] He wanted to remind the scientific specialists of their connections and shared roots with the humanities and to remind the humanists that science and the humanities are merely two facets of the same human endeavour.

The gap between science and humanist culture is undoubtedly

a deep one. It was the main theme of C.P.Snow's influential essay on the schism between the 'two cultures', published in 1959.[15] With the philosophical and political criticism of the scientific–technical rationality that developed in the sixties, the need to 'restore the truly human features in the portrait of science' became urgent.[16] History of science was the natural instrument for this. For example, Sarton's solemn defence of scientific culture is adopted by J.T.Clark who concludes that 'the history of *science* is, in fact and in deed, the *new* humanism for our contemporary, irreversibly technological, and at the present moment beleaguered culture'.[17]

Certainly the study of history of science reveals that the gap between science and humanism is not an inherent feature in Western culture. And certainly history of science can be used to argue that many eminent scientists were and are deeply committed to humanist concerns and that their science contains central human aspects. But such arguments should not be used ideologically, to silence critics of contemporary science. It is, after all, no real argument for the humanity of science that Einstein was a capable violinist or that Oppenheimer wrote poetry and studied Buddhist philosophy. History of science ought, rather, in this connection to be used to ask why much the greater part of our science today can no longer be regarded as an expression of humanist endeavour.

VI. History of science is in no need of pragmatic justification, that is, with reference to contemporary problems. As an important factor in the general cultural and social development, science will naturally attract historical attention in the same way as, for example, religion and economics. Since science has possibly even been the most important factor in the development of modern society, an understanding of history of science is all the more necessary. According to this view, history of science has no particular aim apart from that of revealing the past.

Justifications of this type will typically appeal to historians, but will only have a limited appeal for scientists. Herbert Butterfield reasoned as follows at the beginning of *The Origins of Modern Science*:[18]

> Considering the part played by the sciences in the story of our Western civilization, it is hardly possible to doubt the importance which the history of science will sooner or later acquire both in its own right and as the bridge which has been so long needed between the Arts and the Sciences.

Non-pragmatic history of science, considered as a part of general cultural history, often goes hand in hand with a history-of-science-for-its-own-sake attitude. Many prominent historians are of the opinion that the discipline ought to be cultivated without any kind of external justification. They believe that historiographical standards can be assured and developed best if historians of science write for each other. I.B.Cohen is one of those who have repudiated an externally motivated history of science and have warned colleagues against allowing professional history of science to be dictated by external objectives and demands for relevance. 'At the present time,' he wrote in 1961,[19]

> it is surely no longer necessary to justify the study of history of science. We need seek no 'excuse' for our inquiries into the origin and development of any activity which for more than two millenia has attracted to itself some of the best minds the world has ever known! . . . Not far off is the time when historians of science will be so numerous that they may produce scholarly works which need satisfy only the members of their own profession, the only requirement being that of high standards.

A similar concern for the purity of the discipline was expressed by Pearce Williams in a critical commentary on J.D.Bernal's *Science in History*. This work represents, according to Pearce Williams, a 'majestic myth', an example of how history of science may degrade in quality when it is written with an external objective in view:[20]

> . . . the history of science is a professional and rigorous discipline demanding the same level of skills and scholarship as any other scholarly field. It is time for the scientist to realize that he studies nature and others study him. He is no more nor less competent to comment on his own activities than is the politician and the same is true of the history of science.

In modern history of science, most of the aims referred to above will be acceptable to at least some practitioners. But, because of the proliferation of sub-disciplines and perspectives, no one of the aims can be said to bind the discipline as a whole. The newer developments in history of science, 'the new eclecticism' as they have been called, include a relative decline of purely intellectual history.[21] Historians increasingly attempt to integrate their subjects, intellectual or not, with other historical subjects and methods. New perspectives, inspired by social and economic history in particular, have been incorporated in the discipline. While history of

science has traditionally dealt with the contributions of ind
scientists, today its interests are much broader and generally
towards collective phenomena. Nations, firms, political a₈
institutes and scientific societies are being studied by a rising stream
of historians, many of them being employed by the institutions
they analyse. As to scientific disciplines, physics has traditionally
played a dominant part in history of science. The last decades have
witnessed a vigorous consideration of non-physical sciences,
including the earth sciences, biological sciences, sciences of man
and pseudo-sciences. Whatever science or subject being studied,
history of science is considered by its practitioners increasingly as
a field of history rather than of science.

The problems surrounding the objective and relevance of history
of science are closely connected to the question of the extent to
which we can learn from history. We cannot learn from history
of science how to solve specific scientific problems. But we can
assess and understand our own contemporary science better in its
social context with the aid of knowledge of its history. History of
science provides us with a pool of experience, in which we can
more or less clearly identify trends and relationships. From these
we can learn how to act in order to consolidate or weaken present
tendencies. The fact that we can learn something from history of
science about how better to plan the future of science does not
imply an acceptance of pragmatic historiography of science.

In particular, history of science can give us a useful reminder
that the forms in which science is carried out today are not the
only forms possible but a socially conditioned selection among
many alternatives. Reference to known historical courses can give
us information about which aspects of science are 'natural' or
inherent parts of science *per se*; and, more to the point, information
about which aspects are not but are culturally determined and
therefore part of the social context of contemporary science. It is
history of science in particular which has taught us that the
positivist belief in a value-free, culturally independent science is a
myth. And it is history of science, more than anything else,
philosophy included, that has taught us that *the* scientific method,
perceived as an absolute, canonized doctrine, is an artefact.

Those morals that can be drawn from history will most often
be vague and ambiguous. Typically, instructions in history of science take place on the basis of a collection of case-studies, not all

of which point in the same direction, so that the question of the 'weighing' of conflicting evidence will always arise. Such a weighing cannot be made without reference to theoretical considerations but will involve interpretations that are open to criticism.

As an example, we can take the question of how far a realist or instrumentalist attitude to science is the 'best' or most fruitful one; that is, which of the two attitudes best guarantees progress in scientific knowledge. From the point of view of philosophy, this is an important question. If history is to be able to give an answer, it must do so through a weighed count of which great strides in scientific progress were based on realist and instrumentalist attitudes, respectively. If history unequivocally points out that all great progress has been the work of, for example, scientists of the realist school, while instrumentalists have always had a negative influence, the conclusion would be obvious. Elkana uses this kind of historical argument in his critique of instrumentalism in modern physics:[22]

> It can be shown that, in the past, all those discoveries we admit were steps in progress were made by realists and never by instrumentalists. This applies to the 20th century, too, with the possible exception of Heisenberg. The argument can be fruitful if one takes case history by case history.

The conclusion apart, Elkana's argument is not satisfactory. One will always be able to argue about which scientists were realists and which were instrumentalists; and one will always be able to argue about which discoveries were 'steps in progress'. History itself cannot teach us this. But it can teach us that not *all* great discoveries were made by realists and that instrumentalists have not *always* blocked progress. Rather a trivial lesson.

4

Elements of theory of history

According to one historiographical theory associated with positivism, history is a description of the past, based on a series of well-documented facts. Positivist historiography is based on the following assumptions:

a. History (i.e. the past, H_1) is an objective reality that is the unchangeable object of interest to the historian.

b. It is the task of the historian to reconstruct the past as it actually was, i.e. give a true description of the course of events of the past. But it is not his task to interpret or evaluate the occurrences of the past or to draw conclusions about the present or future on the basis of history. The study of history is the study of the past as the past. This programme can be found in a famous quotation from Ranke:[1]

> To history has been assigned the office of judging the past, of instructing the present for the benefit of future ages. To such high offices this work does not aspire: it wants only to show what actually happened [wie es eigentlich gewesen].

c. It is, in fact, possible to write history 'wie es eigentlich gewesen', i.e. to attain an objective knowledge of parts of the historical past. This epistemological objectivity implies, among other things, that the subject (the historian) can be separated from the object (the historic events) that can be viewed impartially, be seen 'from without'. The ideal of impartiality was expressed by another well-known historian, Lord Acton (1834–1902). In his plan for the collectively authored *Cambridge Modern History* he stressed that impartiality is the hallmark of good historical research, and that the contributors should bear in mind that[2]

> our Waterloo must be one that satisfies French and English, Germans and Dutch alike; that nobody can tell, without examining the list of authors, where the Bishop of Oxford laid down the pen, and whether Fairbairn or Gasquet, Liebermann or Harrison took it up.

d. History can be viewed as an organized sum of simple, particular facts that can be discovered through the study of documents from the past, using methods that are critical of sources. It is the most exalted task of the historian to uncover these facts. Interpretations and conclusions can only be made and drawn when all the relevant facts have been collected. Sarton compared the historian of science with the entomologist.[3] In the one case it is insects that are collected and arranged, in the other it is scientific ideas.

The positivist view of history has little credibility today. Since the end of the last century, many historians have responded sharply to the programme outlined above. Among other things, the naive belief in simple historical facts as the building bricks of history has, with good reason, been problematized.[4] The point here is that a distinction has to be made between 'facts of the past' and 'historical facts'. While the former include everything that actually happened in the past, the latter are the data accepted by the historian as being of such reliability and interest that they appear in historical literature. Only a few of the occurrences of the past achieve 'historical' status. This status is assigned to them by the historian. Historical data as such are not found in the past but they are constructed. Since historical facts are the product of an evaluation and interpretation, they are relative to the interests of the historian. There is no generally accepted criterion for when an occurrence has historical status and can thence enter the arsenal of historical facts.

The relativity of historical data is in accordance with the fact that the facts of the past can be *turned into* historical facts. This is what happens in those cases where previously unnoticed scientific contributions are 'rediscovered'; a classic example is Mendel's discovery of the genetic laws, to be discussed in chapter 9 and chapter 17. Conversely, historical facts can lose their privileged status and fall back into historical oblivion as mere facts about the past. Many discoveries, once considered to be milestones in the progress of science, have turned out to be trivial or erroneous and then, as a result, lost their position in history of science. They are no longer part of the living history.

The historian is thus actively involved in the construction of historical facts. This is the case to a still higher degree with so-called simple historical facts, the basic events (or basic statements) on which positivist historians wished to build history. It is the historian who is interested in describing a fact as simple and who, therefore, artificially isolates a particular episode from a complicated course of events. This is not merely a legitimate but a necessary tactic in the writing of history. Two historians who look at the same historical occurrence but who have different interests will thus not crystallize the same simple historical facts from the occurrence.

Although historical facts do not possess the factual and intersubjective nature often attributed to them, they are not, of course, the haphazard construction of the historian. Almost all historians will agree on the crucial importance of basing history on facts, and not on fantasies, guesses or wishful thinking. But there are differences in the status given to facts in historiography. To the positivist historian, facts are sacred and not to be tampered with and historical writing will often tend to be mere expositions of the facts. Contrary to this, most modern historians regard the accurate revelation of facts as worthless in itself. E.H.Carr has expressed it thus:[5]

> The historian without his facts is rootless and futile; the facts without their historian are dead and meaningless. . . . To praise a historian for his accuracy is like praising an architect for using well-seasoned timber or properly mixed concrete in his building. It is a necessary condition of his work, but not his essential function.

The question of historical facts goes deeper than the question of historical status alone. It also refers to when anything at all is, actually, a fact. According to positivism, pure statements of observation do exist that do not change when there are changes in the theoretical framework and that are therefore unproblematically factual. The English physician and scientist William Gilbert (1544–1603) performed pioneering magnetic experiments which he discussed in a book, *De Magnete*, published in 1600. Admittedly, Gilbert's experiments and interpretations of them were strongly influenced by his theoretical views, his world picture. But, the positivist will assert, it is necessary and also possible to identify Gilbert's immediate observations as facts that are independent of his theory, and it is on these reconstructed facts that history of science must be based. This sharp separation of facts from 'theory' is a fiction, however, according to modern theory of science. Exactly what constitute pure 'facts' and what constitutes 'theory' in Gilbert's reports in *De Magnete* depends on later knowledge of magnetism and will thus be relative to time.[6] The idea widely accepted today about the dependence of observations on theory implies, therefore, that it may be impossible to isolate the empirical facts of the past, in the sense of true statements about nature, uninfluenced by theory. But it does not imply that for that reason history of science must cease to be factual. It is just that the facts sought by the historian are historical, not scientific facts. It is a historical fact, for example, that Gilbert had a particular world picture that caused him to make some magnetic experiments that he described in a particular way.

When the historian has unearthed all the sources he can, he possesses a store of data or facts. These are the product of a selection that have already taken place in the past, since only a very limited part of the events of the past has ever been recorded. In order to turn his data into history, the historian has to make a further selection in accordance with the priorities he wants to make. This selection process forms a constructive or active element that, to some extent, reflects the historian's world view. A number of factors, ranging from personal likes and dislikes to philosophical or political positions, will contribute to a subjectively coloured history. Carr has carried this situation to an extreme by asserting that 'when we take up a work of history, our first concern should be not with the facts which it contains but with the historian who wrote it'.[7] This moral should not, perhaps, be taken literally. But it contains an important kernel of truth that also applies to history of science. Historians who have different world views will naturally select different sources and place the emphasis on different factors and therefore arrive at different conclusions. In such cases it would be instructive to take the views of the historians involved into consideration. In the chapters that follow we shall meet several such cases.

In his critique of facts-fixated positivist historiography, the American historian Charles Beard (1874–1948) examined Ranke's historical practice and demonstrated how this did not in the least comply with what he, Ranke, advocated in his historical programme.[8] On the contrary, retrospectively it can be seen that Ranke's own works clearly express a politically conservative commitment in harmony with the ideology of his age. On the other hand, the fact that historical practice in Ranke and others does not follow their stated programme is not an argument against positivist historiography in particular. It could be used just as well against sceptical historians such as Beard and Becker. Their theoretical scepticism or relativism did not prevent them from working on concrete historical tasks in which they gave definite explanations of historical occurrences and unveiled events as, in their opinion, they actually happened.

The historian's active, socially conditioned, intervention in the historical process explains two important traits in historical works: Firstly, that the same themes and periods are described and explained differently by different historians who, at that, have the same source materials at their disposal. Secondly, that history is always being rewritten. This is only partly due to the fact that new sources are being discovered, forcing such a rewriting to be made. What is more important is that the interpretation of the past is, to some extent, a function of the present. Each new generation of historians looks at the past with new eyes, the eyes of the present. The idea that history, including history of science, is constantly being rewritten was expressed in a precise way by Goethe:[9]

> The history of the world must be rewritten from time to time . . . . But the need to do so does not arise because many things have been discovered, but because new opinions will be created when a person in a later age adopts views from whose vantage point the past may be surveyed and judged in a different way. This is also the case in the sciences.

Among many examples, let us briefly mention one from the history of geology.[10] In this century, James Hutton (1726–1797) has usually been pictured as the true founder of modern geology, a revolutionary and highly original scientist. The other pioneer of geology, Abraham Werner (1749–1817), has been interpreted as an old-fashioned, speculative thinker, if not a counterrevolutionary. However, during most of the 19th century Werner was, at least outside England and Scotland, recognized as the founder of geology while Hutton and his school were only assigned a minor, and not always positive, influence on geology. Histories of geology written in the mid 19th century interpreted the role of Wernerian and Huttonian theories entirely differently, and not necessarily less correctly, than did histories written a century later. Some modern historians of geology argue that Hutton's role has been much exaggerated and that Werner's system was, after all, rational and very important to the development of geology.

Several historians have used the objections to the positivist view of history in support of a sceptical or relativist view of history. The radical sceptic will maintain that we can never achieve any certain knowledge of history, that we know nothing of the past; the radical relativist will maintain that all historical descriptions of the past are equally good or equally bad. We will confine ourselves here to referring to them jointly as sceptical views of history. Beard has given the following version of the sceptical doctrine of history:[11]

> The historian, who writes history, carries out a conscious act of faith as far as order and movement are concerned, since certainty about order and movement through knowledge of that with which he is dealing is denied to him. . . . His belief is, in fact, a conviction that one can have true knowledge of the movement of history, and this conviction is a subjective decision, not an objective revelation.

Beard and Becker both emphasized that historical knowledge has to be indirect, since it concerns events in a past that is, and always will be, inaccessible for direct observation:[12]

> The historian is not an observer of the past that lies beyond his own time. He cannot see it *objectively* as the chemist sees his test tubes and compounds. The historian must 'see' the actuality of history through the medium of documentation. That is his only recourse.

According to Beard, the lack of direct observation means that historical works cannot be submitted to objective tests and thereby be made objects of unequivocal classifications as true or false. Following the same line of thought, Becker maintained that the historical facts with which the historian operates cannot be the actual occurrences of the past, which are beyond observation and manipulation.[13] The historian can only work meaningfully with judgements or statements about these occurrences. It is these retrospective statements alone that constitute historical facts. These are not extracts of actual past reality but are 'symbols' of the reality that is not, itself, accessible to history.

Another argument in support of scepticism touches on the necessarily incomplete nature of historical knowledge. A.C.Danto has connected this fact with the fact that, in a sense, historical statements are always related to the future.[14] Historical occurrences that happened at the time t_0, are always analysed at a later time t_1. In even later times t_2 the occurrence in t_0 will often be described in a completely different way to the way it was described in t_1. Not only because new criteria are being used for selection and evaluation, but also because things have happened between t_1 and

t_2 that give a completely new character to the original occurrence. As an example, Danto gives the statement 'Aristarchus anticipated in 270 BC the theory Copernicus published in AD 1543'. This is an important, true history of science statement; but it could not possibly have been formulated by a historian writing in, say, 1200. In the same way, many important statements about the past can only be formulated in the future and since the future is supposed to be infinite, at any one time there will be an infinity of true, important historical statements that cannot be formulated.

Scepticism is bound up with a so-called *presentist* view of history. According to presentism, the past can never be a goal in itself for the historian who, on the contrary, has to look at the past through the eyes of today and evaluate it critically with the problems of today as his point of departure. History refers not to the past but to the present to whose 'practical needs' it forms a response.[15] It follows that history must and ought to be committed. If not, it will be lifeless cultivation of the past without any meaning or interest. As Nietzsche, who formulated part of presentism's programme for more than a hundred years ago, said, it will be antiquarian instead of critical history.[16]

Presentism has been a popular theory of history in different versions in this century, especially as a result of the Italian philosopher Benedetto Croce's (1866–1952) radical revolt against the positivist view of history. Carr, who sympathizes with presentism concludes as follows:[17]

> The function of the historian is neither to love the past nor to emancipate himself from the past but *to master and understand it as the key to the understanding of the present.*

In Croce and some of his followers, presentism is based on a subjective idealism in which history is purely spiritual. In other authors the presentist view is associated with pragmatic philosophy. This relationship can be seen in presentism's emphasis that history is a means of responding to contemporary problems and is only justified in so far as it can carry out this duty. In view of this it is understandable that the American philosopher John Dewey (1859–1952), one of the fathers of modern pragmatism, made himself spokesman for a presentist view of history.[18] To Dewey history as past is of no interest. History, like science, is an instrument that fulfils practical needs. Historical statements that do this are true.

Outright sceptical or presentist views have not had great penetration in the history of science. But Collingwood's and, to a lesser extent, Croce's views have evoked some response in modern theory of science and history of science. The American philosopher Maurice Finnochiaro has argued that Croce's view of history ought to be especially accepted in history of science. He thinks that Croce's theory lives up to the ideal of the historians that history should be able to be understood intellectually and have an effect on the living individuals that history addresses. If, like Finnochiaro, one accepts Croce's view of history, one is led to the conclusion – the one also drawn by Finnochiaro – that history of science must meet a present need. Since history of science is supposed to be a response to the problems of the scientists, it is the contributions of practising scientists that will be of particular importance to history of science; while the works of professional historians will merely be of antiquarian interest.[19]

It should be pointed out at this stage that Croce's view of history, at least if taken as a whole, is unacceptable as a background for serious history of science; or for any other kind of history, for that matter. If Croce's theory is taken seriously, it is a negation of all history that lays claim to be able to distinguish between true and false historical statements. Similarly, analyses of sources are, to Croce, merely superficial chronicle writing. In principle, the historian can and ought to manage without any sources whatsoever since history, properly, is 'a truth that we have drawn out of our innermost experience'.[20] What historian will be ready to accept that in practice?

Robin Collingwood (1889–1943) who, according to Carr, is the only significant British philosopher of history in this century, has idealism and subjectivism in common with Croce. According to Collingwood, the object of history does not consist of the occurrences of the past but just of the thoughts about these occurrences. The historian must try to re-experience or *re-enact* the thoughts of earlier individuals. When he has succeeded in doing this he knows what has happened and does not need any further information or explanations of why the event took place. 'The history of thought, and therefore all history, is the re-enactment of past thought in the historian's own mind.'[21]

Collingwood's idea of re-enactment covers a special historical form of understanding where the historian has to steep himself in the thoughts of the past and seek a kind of sympathetical harmony with them. Since the understanding acquired in this way can only be valid for actions that result from thought, these are the only kinds of occurrence that form the substance of history. According to Collingwood, science, art and politics belong to this category but not any kind of natural event. This leads to the strange result that biographical descriptions do not belong to history. Collingwood's justification for this is that biographies are structured on biological events – the birth and death of the person concerned – and not on intellectual ones.[22] This result alone ought to provide grounds for scepticism as far as the value of Collingwood's ideas is concerned. Surely it would be artificial to exclude biographies from history.

The really problematical part of Collingwood's view of history lies in the idea of re-enactment and in the assertion of what kind of thoughts can be re-enacted. According to Collingwood, the historian, in order to understand the thoughts of Archimedes, for example, has to rethink these thoughts in his own head:[23]

> We cannot relive the triumph of Archimedes or the bitterness of Marius; but the evidence of what these men thought is in our hands; and in re-enacting these thoughts in our own minds by interpretation of that evidence we can know, so far as there is any knowledge, that the thoughts we create were theirs.

But how is the historian to know that his re-enactment has succeeded, that it is really the thoughts of Archimedes that he is rethinking? Collingwood does not supply any criteria for this but seems to think that it is subjectively determined by the intuition of the historian.

Collingwoodian re-enactment is, in reality, a *rational reconstruction* of the thoughts of the historical agents. Such a reconstruction need not take the authentic thoughts of the agent seriously and is not, therefore, primarily directed at the authentic past.[24] According to Collingwood, it is furthermore not just any intellectual activity that can be re-enacted. This is only the case with reflective, goal-orientated intellectual activity, i.e. the kinds of thoughts that are concerned with the solving of problems.[25]

> In order, therefore, that any particular act of thought should become the subject-matter for history, it must be an act not only of thought but of reflective thought, that is, one which is performed in the consciousness that it is being performed, and is constituted what it is by that consciousness . . . . A reflective activity is one in which we know what it is that we are trying to do, so that when it is done

we know that it is done by seeing that it has conformed to the standard or criterion which was our initial conception of it.

If the historian is to understand the thoughts of Einstein, for example, he must therefore identify the problem at which the thoughts were directed. But the problem is only known via a conclusion reached in reverse from the solution of the problem; this makes Collingwood conclude that we only know, and can only know, the successful thoughts of Einstein, that is, those concerning problems that Einstein actually solved. This leads to the strange result that we can never say a scientist or philosopher concerned himself with problems that he could not solve; for we cannot have any historical knowledge of such problems. This is, of course, an unacceptable result since we do know that Einstein did actually work on problems that he did not solve. All human beings have had failures when trying to solve certain problems and in a great many cases we know about these unresolved problems.

Some parts of Collingwood's programme have had a positive influence on the history of science.[26] This is true of his insistence that historical statements have to be seen as responses to problems and that the historian ought to concentrate on these problems; and it is true of his relativist moral that what is of relevance for history is not a matter of how far past statements were true or false in an absolute sense, but of how they can be understood in the context of a problem; and, finally, of his assertion that the historian ought to attempt to think himself back into the past. But, as with Croce, it would be true to say that Collingwood's historiography, taken as a whole, is unacceptable to history of science. As indicated above, it contains many elements that are in direct conflict with what modern history of science is striving to do.

The arguments presented in support of scepticist historiography contain valuable insights but do not justify scepticism or subjectivism in their strong forms. The sort of weak scepticism that is associated with the problematic nature of the selections and interpretations of sources do not imply that it is pointless to try and distinguish between truth and falsehood in history (see also the following chapter). The radically scepticist or relativist historian asserts, after all, that his own view is true and that his arguments in favour of it are better than arguments against it. Even the sceptic is forced to maintain the truth of some assertions and the falsehood of other.[27]

5

Objectivity in history

The legitimate criticism of the objectivity of historical facts is more concerned with when an occurrence is historical than with when it is a fact. It does not give grounds for doubts as to whether, after all, objectively true statements of fact about the past can actually be established. It is a fact that Caesar crossed the Rubicon in 49 BC and it is a fact that Darwin was born in 1802. Although data like this does not form the kernel of history, the simple establishment of facts is an important element in the process of historical research. The untangling of the facts is potentially valuable even if they cannot be explained at the time or placed in a historical context. The fact that the establishment of data is the result of a process of selection and is probably directed by subjective influences does not make the data less true or less objective. The most it can do is to make them less significant or less interesting. Many historians will regard it as of no consequence that Darwin was born in 1802, but what historian would seriously deny that Darwin was, after all, born in 1802?

When historians are interested in facts about the past it is because of their possible historical status, which, in practice, means their historical significance. We must therefore ask whether there are any objective (in the sense of absolute) criteria for the granting of the epithet 'significant' to some events and 'insignificant' to others. The answer is, with modification, no. In general it is not possible to identify events that are significant in themselves in an absolute sense, that is, across time and place and historical perspective. It will always be possible to see events in a perspective that makes them appear to be unimportant. In more specialist historical writings, however, the freedom to ignore certain events will be limited by the kind of speciality. A history of biology in the 19th century in which Darwin's works are not regarded as important events is almost unimaginable. The objective importance of Darwin's system does not depend on it being regarded as important today or in its having resulted in important modern problems. By virtue alone of the important part that Darwinism played in the science and culture of the late 19th century – an importance that cannot possibly be denied – it has assured itself of such a place in history that no future historian of biology, working according to approximately the same principles as those accepted today, will be able to ignore Darwin's works.

When sceptical historians assert the subjective nature of historical knowledge, it is usually conceived in relation to scientific knowledge with whose reliability and objectivity it is contrasted. In other words, 'objective knowledge' is regarded as being synonymous with 'scientific knowledge'. But facts are not, in themselves, given in science either. Just as in history – though not in precisely the same way as in history – relevant facts are selected by the scientist and these facts often only have a meaning within a specific theoretical framework. Nor is the scientist in a position to unravel 'the whole truth' about the phenomena being studied. Scientific knowledge is incomplete, too, in the sense that it consists of partial truths. There is, therefore, no reason for attributing an especially subjective nature to history because of its selective and incomplete nature. Neither is the fact that historical knowledge is based on sources whose genuineness cannot be rigorously proved, a reason for doubting the quality of that knowledge. Conditions are not essentially different in the empirical sciences which build on observations which, in principle, can always be disputed.

A standard objection to historical objectivity is, as mentioned in the previous chapter, that historical events cannot be directly observed and are inaccessible for testing and for experimental manipulation. Here, too, historical knowledge is placed in opposition to the scientific knowledge that is seen as a paradigm of objectivity. The idea of direct observation as a prerequisite for objective, true knowledge, however, is untenable.[1] It is based on a naive empiricist view of science, a view that has long been shown to be faulty. Scientific knowledge never springs from 'direct observations' but is the product of a process during which observations are selected and evaluated as evidence of varying reliability. The reason why we give an objective ontological status to atomic nuclei is that we have convincing evidence of their existence. This evidence is indirect, based on the study of scattering of radiation on matter, for example. No physicist has ever 'directly observed' an atomic nucleus and it is hardly likely that any will do so in the future. Even so, physicists, and historians, too, do not hesitate to call our knowledge of atomic nuclei objective. In contrast, the zoologist will not accept the existence of the Loch Ness monster as an objective fact, even though this has been 'directly observed' many times.

It is remarkable that many of the most radical antipositivists within the humanist and social sciences cheerfully accept a naive positivist view of the natural sciences. This makes the argument against a scientifically based history, psychology or sociology not only simple but cheap, since what is being argued against, is, in reality, a phantom. I agree with Popper when he maintains that scientism is not an attempt to colonize the humanist and social sciences with the methods of the natural sciences but an attempt to colonize them with what are erroneously believed to be the methods of natural science.[2]

The sceptical historian will perhaps point out that historical events are fundamentally different from those of science, since historical events cannot be repeated and thus cannot be made the object of the experimental manipulation that ensures the objectivity of scientific knowledge. But although they are indeed important, controlled experiments are not absolutely necessary constituents of scientific knowledge. Many of the events studied in science are not manipulatable and repeatable. Most astronomical and geological problems come into this category. Furthermore, the kind of repeatability with which science operates contains a historical element and, in fact, assumes that past events can be recognized objectively. When experiments are repeated and compared, a temporal process is involved, based on a tacit consent of a certain permanence and stability in the world. When scientists carry out a series of similar experiments over a large number of years, a critical comparison can only be meaningful if one believes that the knowledge that stems from the older experiments is still valid. In other words, if one accepts that it is possible to have reliable knowledge about the past.

All in all, the postulated contrast between a present to which there is reliable empirical access, and a past that has been stripped of this quality cannot stand up to closer analysis. As intimated above, this contrast clashes with another of the sceptics' alleged contrasts, the one between objective natural science and non-objective history. If knowledge about the past cannot be objective because it concerns the past, this must not merely apply to the human history of the past but also to the natural history of the past, including palaeontology and large areas of geology and astronomy. Few will be prepared to accept this consequence. There is general agreement that we have objective access to the fauna of the past; that we know that dinosaurs once lived even if they have never been observed directly by human beings.

The historian often draws conclusions on the basis of the evidence of people who are dead, evidence that it is not possible to test using one's own observations. But other historical conclusions are quite independent of eyewitnesses' reports and are of the same kind as standard inductions in the natural sciences. The historian who finds and studies relics of the past is in the same position as the palaeontologist. The historian studies 'clues' or 'tracks' from the past and interprets these; the chemist studies 'clues' and interprets these as the result of molecular changes. The chemist does not manifestly observe the molecules any more than the historian observes the occurrences of the past. As Marc Bloch pointed out:[3]

> ... it is not true that the historian can see what goes on in his laboratory only through the eyes of another person. To be sure, he never arrives until after the experiment has been concluded. But, under favorable circumstances, the experiment leaves behind certain residues which he can see with his own eyes.

Furthermore, in order to uphold the sceptical doctrine, one is forced to introduce an artificial and arbitrary line between an objective contemporary history and a subjective history of the past. The murder of President Kennedy is a historical occurrence that was 'directly observed' and that, therefore, must be assumed to be accessible for historical cognition; while the same is not, according to scepticism, true of the murder of Caesar. But in what respects does our knowledge of the murder of President Kennedy differ, essentially, from our knowledge of the murder of Caesar?

History does not only concern itself with occurrences that could have been, and once were, observable, but which are no longer so because of their location in the past. History also deals with occurrences that cannot be and never could have been observed. This is the case with all courses of events that are not narrowly located in time and space. 'The significance of the French educational system for 19th century science' is a typical historical topic of this kind. A statement like 'science during the Third Republic was inhibited by a centralistic educational system' is a historical statement, the truth of which no single observer in the past would be able to vouch for because of his having been alive in France during the Third Republic. On the contrary, it is a statement that can only be evaluated in the future, in other words, historically. This is in contrast to such a statement as 'Galileo determined the time it takes for bodies to fall from the Tower of Pisa on March 3rd, 1582'. This statement would be able to be verified by a witness who happened to be at the foot of the Tower of Pisa on that particular day.

The kind of objectivity that history possesses is not identical to the objectivity inspired by physics; in the latter sense objectivity means intersubjectively testable knowledge that is completely independent of any human factor. It is only if one clings to the classical rigid definition of objectivity that historical objectivity will have to remain an unattainable ideal. Hooykaas has expressed the dilemma of the historian in his striving for an objectivity that can never be attained:[4]

> What method do we want then? An *objective* one. But objectivity is impossible! Without any doubt, it is impossible, as historiography is not a mere compilation of facts: the choice of material already implies an element of subjectivity and amounts to an evaluation. The fact that the historian of science is a scientist himself, influences his judgment on what is important or not. But in spite of this unavoidable influence of the historian's own political, educational, social, national, religious background and his personal character, we maintain the ideal of objectivity. Like all ideals it is unattainable, but, nevertheless, it should keep us in a holy dissatisfaction with ourselves.

Hooykaas's dilemma rests on the concept of absolute objectivity that reflects an empiricist view of the process of cognition as a passive reception of impulses from outside. This is a view that is not even valid when applied to scientific cognition. There is good reason, therefore, to abandon absolute objectivity and to use more suitable criteria of objectivity that reflect the nature of cognition more accurately.

The fact that the natural sciences are generally regarded as objective is connected with the high degree of consensus and discipline that prevails in scientific communities. In contrast with this, the historiographical arena is obviously beset by discussions about fundamentals and by serious disagreements without any possibility of a neutral evaluation. At least, this is the popular picture of science and history, respectively. If this picture of a conflict-free, objective science is accepted, historical research must, in comparison, appear to be subjective. But in the first place, there is no reason why one should expect to find the same kind of objectivity in the two disciplines. As Blake, examining historical objectivity, concludes: 'We can admit that standards of historical criticism, and therefore what passes these standards, are in constant flux, without conceding this as a ground for questioning whether history can ever be objective.'[5] This is correct, but it needs to be supplemented with the remark that the popular picture of a conflict-free, objective science is an erroneous one on some important counts. The standards of science, too, are in constant flux, though in a less noticeable and less radical way than is the case in history.

Many proposals have been made towards establishing a definition of objectivity appropriate to history. Some authors, including such diverse thinkers as Max Weber and Karl Popper, have maintained a so-called 'perspectivist' viewpoint. The kernel of this is that the formulation of a problem – the questions that are asked, the sources that are selected, the facts that are accepted as being historical, etc. – is subjective and not accessible to rational criticism; but the statements that have been formulated can be evaluated objectively, without it being necessary to accept the perspective that gave rise to them.

According to Popper, the solution to the problem of selection lies in the historian deliberately introducing a pre-conceived, selective framework for whatever interests him.[6] Such frameworks or perspectives will not normally be granted the status of scientific theories, for they cannot be tested. Popper is a perspectivist but not a relativist. He emphasizes that even though one cannot object to the choice of a particular historical perspective, this does not free one from the obligation to work with many different perspectives, that is, to follow a pluralistic method. Nor does it free one from the obligation to investigate all relevant data, whether it fits the point of view or not. According to Popper, the advantage of having a perspectivist viewpoint is that 'we need not worry about all those facts and aspects which have no bearing upon our point of view and which therefore do not interest us'.[7] But how can the historian know in advance which facts have no relevance for his point of view? To declare certain facts irrelevant does not mean that they are irrelevant.

Most historians are of the opinion that historical knowledge is objective in a more comprehensive sense than the perspectivist one. As Schaff, among others, has demonstrated, one can reconcile the activist nature of historical cognition – the fact that the subject actively participates in the historical process – with the thesis that history is an objective process that has actually taken place in the past. Schaff uses the term 'relative objective cognition' for the relativity that is essential for historical objectivity:[8]

> ... only a relative cognition such as this can be objective: for when a particular system of reference has been adopted and a particular research goal agreed on, we receive from this *eo ipso* a criterion for the selection of historical materials that can no longer be arbitrary, subjective, but which has an objective nature because of the given system of reference.

As opposed to in relativism, the truth value of historical statements is not dependent on who formulates the statements, where, when and in what circumstances. The subjective factor cannot be eradicated completely from history. It is an integral part of historical cognition. But the kind of subjectivity under discussion here – the 'good subjectivity' as Schaff calls it – is not incompatible with objective recognition of historical occurrences. 'Bad subjectivity', on the other hand, destroys the credibility of history; it is the subjectivity that stems from the prejudices, personal interests, political sympathies, etc. of the historian. This subjectivity tends to produce ideology instead of knowledge.

Instead of formulating a general criterion of historical objectivity, one can attempt to give reasons why some historical statements, at any rate, are *not* objective. Such reasons cannot be purely normative but must reflect existing historiographical practice, the historians' more or less intuitive notions of when historical statements break with objectivity. One might suggest, for example, that if a historical account X is to be regarded as non-objective, it should at least contain some flaws.[9] If X cannot be criticized at all one will not think of doubting its objectivity. But what faults or flaws will be sufficient to stamp X as non-objective? Not just any flaw; if that were so then no interesting account would be objective. X can contain false statements, but this is neither a necessary nor a sufficient condition for non-objectivity. The flaws that one connects with an account's lack of objectivity are, for example, contradictions, forgeries of sources, deliberately false statements (lies) and prejudiced interpretations. The appearance of such flaws will make an account objectionable but not necessarily non-objective.

Hermerén has suggested that two further conditions need to be satisfied. Firstly, the flaws should make X misleading, that is, tend towards giving a distorted picture of the historical reality. Secondly, the misleading account should be partisan, that is, should favour particular social interests. But partisanship is not in itself a criterion of non-objectivity (the reality can be partisan). Hermerén recommends the following procedure as a practical test of how far a historical account X is non-objective: (1) Examine whether there are any flaws in X. (2) Examine whether these flaws result in X being misleading. (3) Identify the interests or parties involved in the subject dealt with in X. (4) Examine whether X favours one or more of these parties. (5) Examine whether this favourization is due to the fact that X is misleading, i.e. whether X would also favour the parties if X were not misleading.

Undoubtedly this definition of objectivity accords well with historical intuition. But it is not, contrary to what Hermerén believes, a criterion that can be used across different historiographical points of view. Historians holding very different views will perhaps agree that the definition catches non-objectivity, but will nevertheless disagree about the extent to which a particular account is objective. One will always be able to argue that what is a flaw from one point of view is not a flaw from another. The contents of such words as 'distortion', 'misleading' and 'favourization' are by no means invariant across different perspectives. Although in practice one will often achieve agreement about the truth value and objective nature of historical statements, it is hardly possible to establish a usable criterion of objectivity that transcends historiographical differences of opinion. This impossibility is not, however, a trait that is unique to historical research.

Modern historians are very careful when using the term *truth*, which is often felt to be alien to historical research. A historical work will be judged to be inadequate, one-sided or interesting, but rarely true or false. This is partly because true and false are predicates that are only directly valid for statements or conjunctions of statements. Historical accounts do, of course, contain statements about the past but these are bound up with a narrative whole that cannot be broken down into a series of these statements. For example, historical accounts may well be judged to be true even though they contain false statements and, conversely, they will not necessarily be judged to be true even though they consist of true statements. From a logical point of view, the former account will always be false, the latter always true.

Sceptical historiography has recently gained some influence in the history of science via the sociology of science. According to some proponents of the modern relativist programme for sociology and history of science, it is impossible in principle to attain true or objective historical knowledge. In a recent work the arguments put forward are as follows:[10]

> In view of the variety . . . in scientists' accounts of their actions and beliefs, we suggest that it is inappropriate to search for *any* kind of data that can be used to provide a firm bedrock for historical description and analysis. . . . only God may be able to discern the historical reality lying behind the diversity of actors' versions. As mere humans, we have to accept that both historians and participants can produce many different historical accounts and ground them in evidence.

So, according to this version of relativism, it is not the task of the historian to unravel what really did happen in the past. He can only reproduce and reflect on the accounts that scientists supply about their works. This kind of relativism is based on the belief that the scientists' own accounts are the ultimate raw materials of the history of science. But this is a belief that cannot stand up to criticism. As we shall see later (chapter 13), the historian is often able to uncover truths that go against or beyond the scientists' own accounts.

The strong relativist programme implies, furthermore, that the creative acts of scientists, the classical focus of intellectual history, are beyond historical or philosophical analysis. The process of creation is something with which the historian should not be concerned:[11]

> . . . we, like Popper, are not concerned with the context of discovery, or rather creativity; this we are prepared to accept as a 'black

box'; . . . We are concerned with the processes of acceptance and rejection of beliefs, a process which starts as soon as an idea becomes an action.

But certainly the creative act can be analysed. Historians need not accept creativity as a 'black box' and neither need concern with scientific creativity exclude concern with 'processes of acceptance and rejection of beliefs'. The practice of history of science testifies to this every day.

6

Explanations

A very considerable part of history of science is descriptive, that is, accounts of what occurrences took place and when they happened. In spite of this practice, almost all historians agree that history ought also to be explanatory. A pure description of the past will not qualify as real history but is what is somewhat condescendingly called chronicle writing.

Obviously, not all occurrences are in need of an explanation. In particular, it is the novel, non-trivial occurrences that we want to explain by grounding them in relatively familiar and known experiences. In the first instance, scientific occurrences ought to be evaluated and explained in accordance with the norm or norms prevailing at the time they took place. A period's norm can be regarded as everything that is taken for granted by the scientific community during that period. In this respect, the identification of the norms of a period is important. According to David Knight, 'the recovery of the norm [is] itself interesting and must be the primary task of the historian'.[1]

When a particular norm has been identified it can in itself constitute a basis for explanation. If we ask why a theory was accepted or why an experiment was interpreted in a particular way, a reference to the fact that it was in agreement with prevailing standards can in itself be an explanation. In contrast, a norm-breaking occurrence needs an explanation of its own. The norms that are used as a basis for explanation in such cases should, of course, be the norms of that time, not ours.

There are two main types of suggestion for what should count as an historical explanation. One group has been inspired by the causal explanations used in the natural sciences while the other group stresses more explicitly historical forms of explanation in terms of motives, reasons, understanding, and so on. In practice, many historians operate with explanations as though to explain why something happened is to state its cause. This identification stands out in the following quotation from Samuel Lilley:[2]

> Any historical study must necessarily pass through two stages. In the first events are chronicled – the important point is to discover exactly what happened, in a descriptive sense, and exactly when. When sufficient chronicling has been done, the second stage is reached – the problem now is to establish causal relationships between events, to come to understand *why* things happened as they did.

A causal model much discussed is that suggested by Carl Hempel in 1942 according to which historical events are to be explained nomologically and deductively.[3] Hempel's model is also known as the DN (deductive–nomological) model or the 'covering law theory'. It is bound up with the positivist ideal of science since it carries the pattern of explanation in natural science over into the social and historical sciences. The principle behind explanations according to Hempel is as follows: if an event X (explanandum) is to be explainable, X must be able to be deduced from two sets of preceding premises that constitute what explain (explanans). These are a series of data $c_1, c_2, \ldots, c_n$ (antecedent conditions) and a series of general laws $L_1, L_2, \ldots, L_m$ that cover X and c_i. In formal language:

$$c_1, c_2, \ldots, c_n$$
$$\underline{L_1, L_2, \ldots, L_m}$$
therefore: X

A typical explanation of this kind can be answering the question 'why does this particular cannon ball hit this particular place?'. In this case, c_i are the starting and boundary conditions (the elevation of the cannon barrel, the mass and starting speed of the missile, etc.), while L_i are the laws of mechanics. Once we have specified c_i and L_i we will have explained the event. It is worth noting that the logical structure of the DN model involves that explanation is equivalent with prediction. If one knew c_i and L_i before X happened, one would be able to predict X which would count as an explanation too.

Hempel and other advocates of the DN scheme regard it as a normative model of explanation, not a model that describes all or just most explanations as actually used in practice. As far as history is concerned, very few explanations accord with the DN model, in part because covering laws are not normally formulated in history. It has been maintained, however, in defence of the relevance of the model, that such laws are, in fact, used in standard historical explanations, but they are implicit. When they are not formulated it is, for example, because they are trivial statements about how people normally or always behave. Such lawlike statements may be statements of how any rational person will behave in a particular situation.

Consider the question, 'why did Kepler think that the orbit of Mars is elliptical (and not circular)?'.[4] According to the DN model, an explanation can formally be stated as follows:

$c_1 - c_{n-1}$	:	Kepler thought that the $R_1, R_2, \ldots, R_{n-1}$ were true.
c_n	:	Kepler was a rationally thinking human being.
L	:	Any rationally thinking human being who accepts $R_1, R_2, \ldots, R_{n-1}$ as true will conclude that p is valid.

Therefore X: Kepler thought that p is valid.

Here p stands for 'the orbit of Mars is elliptical', while the statements $R_1, R_2, \ldots, R_{n-1}$ refer to Kepler's knowledge or beliefs concerning Mars, including Tycho Brahe's empirical data.

But the suggested DN explanation of Kepler's idea cannot be regarded as a satisfactory historical explanation. What is of interest to the historian is why Kepler discovered the form of the orbit of Mars while other scientists did not. Such an explanation, if framed in the DN manner, would have to refer to Kepler's person in L. But then it cannot be based on a general law.

Many scientific occurrences are the result of the idiosyncrasies of a particular scientist and cannot be grounded in actions that are valid for 'any rational human being'. For example, Kepler believed that there were necessarily six planets and that their distances from the sun could be related to the geometrical structure of the five regular polyhedra. Kepler's contemporaries, to the extent they followed the Copernican system of the world, thought, as he did, that there were six planets; but the linking with the regular polyhedra was peculiar to Kepler and did not win support among other 'rational human beings'. On the contrary, it was regarded as speculative and unreasonable. So in the case of Kepler's theory one cannot use a covering law about rational behaviour. One has to use a formulation in L that refers to Kepler's idiosyncratic views.

Since it lies in the very nature of discoveries that they are creative innovations, not logical conclusions from data that 'any rationally thinking human being' might have drawn, it does not seem possible to explain discoveries deductively in accordance with the DN model. Scientific discoveries can be understood, explained and analysed so that they appear to be well founded and reasonable in the circumstances. But the DN model is not suited to this purpose:[5]

> To say that satisfactory explanations of scientific discoveries are law-covered is to suggest that the cognitive behavior involved in scientific behavior is rule-governed behavior; but rule-governed behaviour is not creative, whereas it is of the essence of discovering that it be creative: hence if scientific discoveries are to involve creative behavior then it is necessary to explain them as instances of rule breaking behavior; that is, the explanations of them should be law-free.

The fundamental flaw of the DN model is that its explanations do not lead to insight or understanding. Many historians maintain that when an action has been made understandable it has thereby also been explained. If the reader experiences a historical narrative as 'making sense', so that he understands what happened, a rapport between description and explanation will have been ensured and he will not feel any need for further explanation. 'Every [good] historical narrative is, in an appropriate sense, self-explanatory,' writes Gallie.[6] To understand a human action will typically consist of unravelling the intention behind it or in stating the motives or reasons that led the historical agent to act as he did. In such cases one can talk of explanations by intention or reason. These explanations differ from explanations of the Hempel type in that they are not nomological.

In the explanation-as-understanding concept is implied the attractive assumption that in practice we will not regard an explanation of an action as acceptable unless we have also understood it. Another attractive feature is that explanations, according to this view, may consist in merely stating the circumstances that made an action possible, without them being necessary. The question 'why did X discover P in the year t?' will, according to the DN model, require reasons from which the explanandum follows by necessity. But most historians will think that in the same circumstances X could have failed to discover P in the year t; or that some other person could have discovered P in the year t, instead.

Explanations by reason, in order to be objective, must assume that one can reach agreement about when an action 'makes sense'. Since they appeal to a common knowledge of, or feeling about, when an action is rational they are often called *rationality explanations*. An example of a rationality explanation is supplied by N.Roll–Hansen. At the beginning of this century biologists discussed whether evolutionary variation was continuous or discontinuous. A heated controversy took place between two groups, the Mendelians and the biometricians. After some years of dispute, the majority of biologists adopted the answer supplied by the Mendelians. Why? Roll–Hansen argues that this occurrence can be explained by showing that the biologists had good scientific reasons for acting as they did: 'There is no need to involve either psychological or sociological factors in explaining the biologists' preference for Mendelism. In the rationalist view, it simply emerged as the best supported theory according to generally accepted methodological rules.'[7]

W.H.Dray has developed the concept of rationality explanation as an alternative to the DN model.[8] According to Dray, a historical explanation is a normative reconstruction of an action, constructed in such a way that the rationality in the way the historical agent behaved is evaluated. If the action of the agent is shown to be rational, if it is demonstrated that he had 'good reasons' to do as he did, the explanation has succeeded. If it should be shown that the action is notoriously irrational this will make it inexplicable. Dray belongs to the large group of theorists of history who, though they have different views in general, see the historian's task as that of establishing a rational reconstruction of the historical occurrence. Popper, Laudan, Lakatos and Collingwood belong to this group, as does Hempel.

An example may illustrate the nature of the rationality explanation.[9] Let us suppose that a biologist, Jones, accepted the Mendelian theory of inheritance while rejecting that of the biometricians (cf. above). The historian would like to know why Jones accepted Mendelism. An explanation could be 'because there was overwhelming empirical support of the Mendelian theory'. But if the question is answered with 'because there was overwhelming empirical evidence *against* the Mendelian theory', one will feel that the answer is not an explanation at all, although it *could* be true. If there was overwhelming empirical evidence against Mendelism at the time when Jones supported it, then we would expect, as long as Jones is assumed to be a normal, rational person, that it was for other reasons that Jones supported Mendelism.

The concept of rationality is itself subject to historical and cultural changes. What kind of rationality should one count on in historical explanations? The one that the historian regards as rational (the present) or the one regarded as rational by the historical agent? Could one use such laws as 'every rational agent accepts theories when there is strong empirical support for them' about periods in which this law was not, in fact, accepted? The question deals with the extent to which the historian should accept the actual behaviour of historical agents and the actual course of events, as revealed by the best possible sources, as explanations; or whether he should perhaps criticize the agent for not having acted with sufficient rationality, in other words ask whether the agent's motives were well founded in relation to the norm of rationality of that time (or of the present). With respect to the history of science and ideas Laudan has answered as follows:[10]

> ... if we can show that a thinker accepted a certain belief which was really the best available in the situation, then we feel that our explanatory task is over. Implicit in this way of looking at the matter is the assumption that *when a thinker does what it is rational to do, we need inquire no further into the causes of his action*; whereas, when he does what is in fact irrational – even if he believes it to be rational – we require some further explanation.

Thus, according to Laudan it is not necessarily the historical agent's own view of what is rational that should serve as a basis of explanation. Laudan seems to assume that absolute criteria of when actions are 'in fact' rational do exist, and that they should therefore be used. But an absolute criterion of rationality that harmonizes with the lessons of history of science does not exist. If modern standards of rationality are used in evaluating historical occurrences, it will almost surely lead to anachronisms.

Usually historical explanations are thought of as answers to why-questions. Finnochiaro criticizes this assumption and argues that explanations, as far as they refer to discoveries, are answers to why-not questions.[11] Instead of asking 'why did Galileo discover the parabolic orbit of projectiles?' one should, according to Finnochiaro, ask 'why was the parabolic motion of projectiles not discovered before Galileo?'. The two questions are not equivalent. The historian who addresses the first question will have to give reasons why Galileo did, in fact, make the discovery. The historian who wishes to answer the second question will have to state the circumstances that prevented earlier scientists from making the discovery. The why-not question is the most interesting and the one we are most curious to know the answer to. In a sense, we feel that the parabolic orbit of projectiles *ought* to have been discovered generations before Galileo.

Many discoveries, however, are of such a kind that what we marvel at are why they were made, not why they were not made earlier. These are the unexpected discoveries. When Ramsay and Rayleigh discovered the inert gas argon in 1894 it took the scientific community with surprise since there were no theoretical grounds for supposing the existence of new elements in the atmosphere. On the contrary, there were good theoretical grounds, such as the periodic system, against the new element. It would not therefore be fruitful to ask 'why was not argon discovered before 1894?' while 'why was argon discovered in 1894?' is an interesting question.

History of science must be able to account for pseudo-discoveries in the same way as it accounts for discoveries that are accepted at the moment. Historically, cognitively and socially, pseudo-discoveries do not differ significantly from discoveries.[12] What we now call pseudo-discoveries were once accepted as discoveries by the scientific community or by parts of it. Pseudo-discoveries from the last century include such examples as N-rays, J-rays, the Piltdown man, Kammerer's frog, *Bathybus haeckelii*, sub-electrons as well as false chemical elements.[13]

The thoroughly investigated case of N-rays may serve as an example of a pseudo-discovery.[14] In 1903 the French physicist René Blondlot (1849–1930) reported the existence of a new species of radiation which he called N-rays; in the years that followed, the properties of the radiation were studied by many scientists. However, in about 1908 it was concluded that the N-rays did not exist after all. Still there is nothing in Blondlot's pseudo-discovery that significantly differs from more respectable, contemporary discoveries such as those of X-rays and radioactivity. But knowing as we do that N-rays are a pseudo-phenomenon, it would seem absurd to ask why they were not discovered before 1903. What does surprise us, and what we will want to explain, is the fact that N-rays were discovered at all. An explanation of this must necessarily contain sociological elements. 'Why were N-rays discovered?' means, in fact, 'why were N-rays granted the status of a discovery?'. Brannigan regards scientific discoveries as characterized by their social status, which leads to a very different kind of explanation to the traditional one. 'The occurrence of discoveries in a culture must be viewed, not from the naturalistic question of what made them happen, but, following Winch and Wittgenstein, from the question of how they were identified as discoveries.'[15] The question 'why was X granted the status of the discoverer of P' must reasonably be answered in quite a different way to the question 'why did X discover P?'.

The occurrences that one may wish to explain in the history of science are varied in kind and hence cannot be explained in quite the same way. In particular, one should distinguish between individual and collective occurrences where the latter are those that include a large number of people and usually take place over a long period of time. While 'Harvey discovered the circulation of the blood' is an individual occurrence, 'the industrial revolution in England' is a collective occurrence.

Several authors, including Popper, J.W. Watkins and F.A.Hayek, have argued that 'we shall not have arrived at rock-bottom explanations of ... large-scale phenomena until we have deduced an account of them from statements about the dispositions, beliefs, resources, and inter-relations of individuals'.[16] Consequently, 'holistic' explanations in terms of rules that go beyond the individual do not have any justification. This viewpoint is called methodological individualism. According to this doctrine, explanations that are based on 'spirit of age', 'class struggle', 'social interests' or 'intellectual environment' are pseudo-explanations. According to Popper methodological individualism states that 'we must try to understand all collective phenomena as due to actions, interactions, aims, hopes and thoughts of individual men, and as due to traditions created and preserved by individual men'.[17] This does not imply that the historian who wants to explain the industrial revolution is forced to investigate the actions, ideas and motives of each individual participating in the industrial revolution; he is justified in basing (and in practice forced to base) his explanation on the actions, ideas and motives of 'ideal' or 'anonymous' individuals.

One of the arguments in favour of methodological individualism is that one has to base an explanation on something that is known; and this is interpreted as that which can be directly observed. Only quantities that are directly observable can be used as a basis of explanation. 'A theoretical understanding of an abstract social structure should be derived from more empirical beliefs about concrete individuals', says Watkins.[18] This is another version of the myth of direct observability which, as we have seen (chapter 5), is objectionable. The doctrine is especially unacceptable for historical explanations, since one does not even have empirical access to individual historical phenomena. And even if they could be observed, as in parts of contemporary history, it does not make them directly observable; the identification of individual phenomena always involves interpretations.

The reductionist, methodological individualism of Popper, Watkins and Hayek contain valuable traits on a negative level, as a critique of facile holism and social determinism; but it should be dismissed as a requirement for historical explanations in general. Many collective phenomena just cannot be reduced to individual phenomena and cannot be explained on a purely individualistic basis. If one limits historical explanations to explanations based on individual, empirically provable quantities one will merely end up by explaining fewer phenomena, and by explaining them worse.

7

Hypothetical history

Because of their placement in the past, historical occurrences cannot be re-created or manipulated. For this reason hypothetical or contrary-to-fact statements are often regarded as unacceptable in historical works. Thus Joseph Needham says: 'Whether a given fact would have got itself discovered by some other person than the historical discoverer had he not lived it is certainly profitless and probably meaningless to enquire.'[1]

A contrary-to-fact statement is a statement based on an assumption that is known to be factually false, in other words, that cannot be reconciled with the known facts. Such statements are also called counterfactual statements. They contain the conditional 'if' followed by the false statement P. 'If X had not been the case, Y would not have taken place' is a counterfactual statement in so far as X actually was the case (irrespective of whether Y occurred or not). X might, for example, be 'Maxwell formulated the theory of electrodynamics' and Y might be 'the radio was invented'. In a certain sense the statement can be said to be a hypothetical statement about the past; but with the difference that the premise of the hypothesis (non-X) is known to be false. Hypotheses are normally statements whose truth value is not known, but which are used heuristically in order to deduce testable statements that will then support or weaken the hypothesis.

We cannot know whether the radio would have been discovered had Maxwell never lived; for we cannot remake the historical situation at the time of Maxwell without taking into consideration the fact that Maxwell did actually live. Counterfactual history seems to presuppose that individual historical occurrences can be taken out of their context without disturbing anything more than a few other occurrences. According to many historians with a 'holistic' view, this presupposition is fundamentally unjustified, since all historical occurrences are connected to each other. The assumption that an actual occurrence had not taken place would have changed all subsequent occurrences in a totally unpredictable way.

In spite of these objections and in spite of the fact that we can never determine the truth value of counterfactual historical situations with certainty (but see below), they are of value in history. In practice, counterfactual questions are not infrequent in the history of science. According to Bernal, 'we ought to demand not only how was this discovery made, but why was it not made before then and what would have been the course if history had gone differently'.[2] Bernal continues by supplying an example of such a counterfactual scenario:[3]

> For instance, if the casual conversation of Henri Poincaré and Bequerel in 1897 had not taken place, it might have been many years before radioactivity was discovered. Ultimately, it was bound to come, because there are many effects which are traceable to it, but it would have been much harder to interpret. If the discovery of radioactivity had been delayed the result of human history might have been quite different. The Second World War and atomic fission only came together in time by the merest accident. If the bomb had come four years earlier we should have had atom bombs fully in use all through the war . . .

Questions of why occurrences took place as they did are of course an important part of history. Such factual questions can, however, also be formulated counterfactually, especially when they are attempts to make causal connections between occurrences. If A and B are real occurrences 'B was caused by A' can also be formulated 'if non-A, then non-B', which is a counterfactual statement. As a general rule, explanations based on laws contain counterfactual statements.[4] To dismiss counterfactual historiography would be the same as denying the legitimacy of law based explanations.

When one asserts that Maxwell's electrodynamics was one of the causes of, or an important precondition for, the invention of the radio, it is a statement that refers to actual occurrences. If electrodynamics had not been formulated in the way Maxwell did, the radio would not have been invented; or, perhaps, the history of the radio would have run a different course to the one it did. The two versions of the counterfactual statement are not equivalent. 'If non-X, then non-Y' is the same as saying that X is a necessary condition for Y and is thus a rather strong assertion. 'If X had not occurred then Y would not have occurred in the way it did' is a weaker, but often more reasonable assertion. It does not exclude the possibility that Y might nevertheless have occurred, perhaps later and in a different form.

Hypothetical historiography can take place with the use of *ceteris paribus* clauses, in other words by using an all-else-being-equal assumption. The idea is to stabilize all factors other than the ones being studied, so that these relationships can be studied in isolation, without the 'disturbance' that would otherwise result from the other factors. Arguments of the *ceteris paribus* type are counterfactual, since all else is not equal. A particular historical occurrence X will influence a lot of other occurrences in an incalculable, largely unknown way. If we want to study how far X was the cause of Y we can assume that all these other occurrences were uninfluenced by X. In that case we would not study the real history, but a hypothetical history.

One has to be careful about forming conclusions from counterfactual situations. For example, it is not quite true that 'if non-A, then non-B' always implies that 'the reason for B was A'. There are several reasons why we cannot always find *the* reason for B with the help of counterfactual statements.[5] In order to assess the validity of 'if non-A, then non-B' one operates with a *ceteris paribus* clause to the effect that the world would have been the same if A had not taken place. For example, we must assume non-C, where C represents the occurrences that have A as a necessary cause. And we must assume non-D, where D represents the causes that are sufficient for A. Since we do not normally have a very good knowledge of causal relationships in the past, we cannot know which Cs and Ds are involved and what consequences this has for other occurrences.

Even if one was able to establish the validity of the statement 'if non-A, then non-B', it would still not be enough to establish A as the sole cause of B. Historical causal relationships are usually 'weak', where 'if A, then B' does not rule out the possibility of other causes of B than A. In such cases as these, the statement 'if non-A, then non-B' is not equivalent with 'A was the cause of B'. 'If Ptolemy had possessed a good telescope, he would not have created his astronomical system' is a reasonable historical statement, formulated counterfactually; but it does not imply that 'the reason why Ptolemy created his astronomical system was that he did not have a good telescope'. To an historian, the latter statement is nonsense.

Hypothetical history of science statements can be a means of pointing out the dependence or independence of different research programmes. For example, Kuhn writes that Einstein's early research programme was 'a program so nearly independent of Planck's that it would almost certainly have led to the black-body law even if Planck had never lived'.[6] This is a counterfactual statement, since Planck did actually live and Einstein did not discover the black-body law. What Kuhn wishes to emphasize is that the entire logic behind Einstein's programme was of such a kind that it was independent of Planck's programme but nevertheless pointed towards the result that was first achieved by Planck. Neither Kuhn nor others can, of course, know with certainty whether the black-body law would have been discovered without Planck. But from knowledge of the early works of Einstein good arguments can be found for claiming that, all other things being equal, this would have happened.

The situation would be quite different if Einstein had actually discovered the law independently of Planck; for example, one might come across an unpublished Einstein manuscript showing that he did. Then it would be a case of a simultaneous, independent discovery and Kuhn's reasoned conjecture would become historical knowledge. Counterfactual questions of the type that have been mentioned can thus in some cases be answered with a firm 'yes'; viz., in cases of simultaneous, independent discoveries. If A_1 discovered X at the time t and A_2 discovered X at approximately the same time, but independently of A_1, one can conclude that X would have been discovered at about the time t even if A_1 had never lived. It is obvious that this possibility only applies to affirmative answers. 'If A_1 had never lived, X would not have happened' cannot be confirmed in the same way. Either A_1 was the only one who discovered X, in which case it is a general, open counterfactual statement; or A_2 simultaneously and independently discovered X, in which case the statement is false; or X was not discovered at all, in which case the statement is meaningless.

Georges Canguilhem has demonstrated the untenability of rejecting hypothetical historiography in connection with an evaluation of the historical significance of Gregor Mendel, the founder of genetics. What would the history of biology have looked like had the monastery in Brünn, with Mendel's theses and papers, burnt down in 1865? If this had happened, and if Mendel's published works had also disappeared, Mendel would never have been 'rediscovered'. Since his influence would then presumably have been zero, Mendel would not appear at all in the history of biology. Such a history would be hypothetical today but was quite realistic at the end of the 19th century. It was not until after 1900, when Mendel's laws had been rediscovered, that a historian could regard Mendel as an important figure in the development of biology.

Admittedly, a form of anachronism is contained in attempts to answer the counterfactual question of what the development of biology would have looked like if Mendel had been more highly thought of by his contemporaries. On the other hand, one would feel it unnatural not to ask the question and to pretend that Mendel was not an important figure in the history of 19th century biology. Canguilhem gives the following defence of hypothetical historiography:[7]

> In whose name could anyone demand that such a temptation [concerning Mendel's hypothetical role] should be resisted? No historian, whatever school he may belong to, rejects the possibility that he might understand what has been by imagining what might have been and does so by either thinking up or, conversely, removing causal factors. The imaginary construction of a possible future does not follow from an attempt to deny the past its actual course. On the contrary, it stresses the true historical nature of the past in its relationship to man's responsibility, whether it be that of the scientists or that of the politicians; it purges the historical account of everything that might resemble a dictate of fate.

8

Structure and organization

The structural framework of the historian includes, among other things, divisions into historical periods. Obviously, periodization is the work of the historian, not of history. No objective or natural way of dividing up is to be found inbuilt into the historical course of events. This does not mean, however, that all ways of organizing the historical materials are equally good. In the historiography of modern science a tradition has arisen for working with chronological periods that follow the century in question: science in the 20th century, in the 19th, 18th and 17th centuries. The division is obviously arbitrary, in the sense that it does not reflect any internal tendency in the development of science. By chance, it would be reasonable to distinguish between the 19th and 20th centuries in the history of physics, whereas this is not the case in the history of biology or in the history of the earth sciences.

The periods used will normally be chronological so that the development is simply followed through linear time. But one does not have to regard chronologically simultaneous occurrences as being historically simultaneous too. For example, one could decide to place occurrences into periods according to their more or less natural connection in the hope that this would reflect the internal or logical development of science. If so, scientists who were 'ahead of their time' may be moved to the chronologically later periods to which they are thought to belong naturally.[1] However, this kind of division into periods can easily lead to an anachronistic history, a fable about how the development could and perhaps ought to have occurred. If one regards the atomic theory of Democritus as historically simultaneous with Boyle, or Leonardo's ideas on aviation as simultaneous with the Lilienthal brothers, one cannot avoid forcing the actual course of development. Chronological, linear time is, after all, the natural frame of reference for historical periodization. The causal connection between occurrences goes from the past to the present, i.e. follows linear time. Division into periods in accordance with other parameters than normal time can have didactic advantages but ought to be used with caution. To the few authors who use a cyclic concept of time, historical and chronological simultaneity are not necessarily identical. For example, Oswald Spengler asserted, on the basis of his cyclic view of time and history, that Descartes was 'simultaneous' with Heraclitus, since they represented similar phases in the cycles of two cultures.[2]

The author of a comprehensive work on the development of science, or of a particular discipline, will have to face the question of what emphasis (in practice: how many pages) should be given to different periods of time. The answer given to this question involves a historiographical choice. No periods exist that are more interesting than others as such, i.e. independently of theoretical considerations. In some histories of science the Middle Ages scarcely appears at all, while in other histories the Middle Ages takes up a dominant place; without one being able to say that the one allocation of priority is inherently better than the other. The question of what weight should be placed on different periods was of relevance to Sarton, Bernal, Singer, Wolf and others who wrote comprehensive histories covering a wide span of time.[3] But today the belief in the existence of a natural allocation of priority to certain themes or periods has been abandoned.

Historians of science have discussed to what extent the so-called Scientific Revolution is real or not, i.e. whether there was a natural, historical period from Copernicus to Newton during which natural philosophy was transformed into modern science. Since Duhem, several historians have maintained that the Scientific Revolution is an illusion, since all the elements that are normally associated with good science can be found as early as the late Middle Ages. According to Duhem, the 17th century is not a particularly interesting or revolutionary period. It was just the provisional culmination of an evolution that had its roots a long way back in the Middle Ages. A.C.Crombie, a specialist on the Middle Ages in the tradition of Duhem, has expressed this attitude as follows:[4]

> Modern science owes most of its success to the use of the inductive and experimental procedures, constituting what is often called 'the experimental method'. . . . the modern, systematic understanding of at least the qualitative aspect of this method was created by the

philosophers of the West in the thirteenth century. It was they who tranformed the Greek geometrical method into the experimental science of the modern world.

In contrast, other historians (Koyré, Hall, Butterfield, and others) have regarded the 17th century as a true revolutionary period, the scientific century par excellence. For that reason, they have divided up the history in relation to that period. Koyré had this to say on the subject:[5]

> ... the apparent continuity in the development of medieval and modern physics (a continuity so emphatically stressed by Caverni and Duhem) is an illusion. It is true, of course, that an unbroken tradition leads from the works of the Parisian Nominalists to those of Benedetti, Bruno, Galileo and Descartes. ... Still the conclusion drawn therefrom by Duhem is a delusion: a well-prepared revolution is nevertheless a revolution ....

The fact that Crombie and Koyré can have such different views of the Scientific Revolution is reasoned in their different views of what characterizes modern science. If critical approach, experimental and logical techniques (induction and deduction) and practical orientation are seen as the essence of science, one is led with Crombie to an evolutionary view. In that case, the term the Scientific Revolution is merely an empty label. Koyré's view of science, on the other hand, is different and so, accordingly, is his periodization. The essence of science, according to Koyré, is the application of mathematical methods in the study of nature and the belief that mathematically based theory has priority over experience. The perspective of theory of science decides the extent to which the Scientific Revolution is a real phenomenon or not. In a social history perspective, represented by neither Crombie nor Koyré, the characteristic feature of modern science is its institutionalization and social structure. It will be natural to regard the 17th century as a revolutionary period from this perspective, but from different motives to those of Koyré.

Division into periods and historical labels are usually retrospective in the sense that they are not to be found during the period in question. As a principle of organization for history of science the Scientific Revolution is largely a child of the present century.[6] But the historian is, of course, within his rights in organizing his materials independently of views held in the past. Periodization expresses an evaluation of a whole that embraces past, present and future. In many ways the often heated discussion about the 'reality' of the Scientific Revolution may seem uninteresting. As long as one recognizes that such questions depend on perspective, and as long as one avoids the primitive Victorian version of the 17th century as the sudden time of birth of science, it matters little whether one calls this period a revolution or not.

History of science encompasses a large number of different, individual sciences. What emphasis should a general history of science place on astronomy, for example, compared with anatomy? It is again true to say that there is no natural or objective answer to this question. Perhaps astronomy developed faster and aroused more interest than anatomy in a certain period. But this does not imply that the historian has to use more space on astronomy than on anatomy. He can legitimately maintain that from his perspective of the history of science, anatomy is the more interesting and hence deserves a more extensive treatment.

Traditionally, history of science has been dominated by the physical sciences and, to a somewhat less extent, biology. Although recently less glamorous sciences, like geology, have also met with increasing historical interest, the preoccupation with physics continues. There is no objective reason why geology should occupy an inferior position to that of physics in the history of science. But that is the way it is. In a broader perspective the concentration on physics and biology is certainly unfortunate. As Mott Greene comments:[7]

> It seems . . . to border on irresponsibility to allow a student to write the nth thesis on the reception of Darwinism somewhere, or the $n+1$ thesis on Newtonian minutiae, while an entire branch of science (and, moreover, one with similarly engrossing philosophical implications) is all but unexplored. In such a field, a student would have the chance to produce original and valuable historical work, rather than toting bones from one graveyard to another.

'Success' seems to be the main criterion of historical importance, in the sense that successful scientific work is usually given high priority. A scientist may be judged successful either because his work has turned out to be important for later developments, or because he was a dominating figure in the science of his time. In the first case historical importance is associated with scientific truth, in the latter with a specific social context. The two criteria should not be confused. Historical significance cannot be understood abstractly or logically. Whatever association there is between historical significance and scientific truth value is a matter of contingency.

When Lyell is considered a very important figure in the history of geology it is mainly, but not only, because his uniformitarian system contained a fundamental truth as compared with earlier doctrines: that the geological history of the earth can be explained by slow, natural processes still proceeding today. The work of Lyell's opponent L. Elie de Beaumont (1798–1874) was not pioneering in the same sense but still Elie de Beaumont ranks high in modern histories of geology. True or false, his views exerted an enormous influence on the geological community. 'In spite of their oddity from the modern point of view,' writes Greene, 'his theories must be recognized to have asserted a profound and lasting influence on the development of geotectonics – an influence far greater than that of Lyell.'[8]

The focusing on successful pioneer work may lead to anachronistic historiography. In addition, it may lead to problems in connection with the history of contemporary science. Modern scientific occurrences, because of their short life span to date, can seldom be assessed with regard to whether they are really pioneering or not. Imagine a historian who, in 1690, wished to write a history of contemporary physics. If he based his work on the criterion that only successful and epoch-making discoveries should be given high priority, probably Newton's *Principia* would figure only insignificantly. Needless to say, a historian writing 100 years later would judge the merits of *Principia* in a very different way.

The creation of great scientific ideas has traditionally been the main concern of historians of science. Who first created this or that theory? How and when did the theory come into being? Questions like these have dominated history of science for a long time with the result that the development of science was pictured as a sequence of great discoveries.[9] Implicitly the picture assumes that discoveries are sufficient in themselves to account for the growth of science; that they automatically enter the corpus of scientific knowledge once they have been made. However, important discoveries are, in fact, often ignored or rejected. Discoveries rarely act instantly. There is, therefore, good reason to call attention to the importance of the diffusion and further development of the ideas that result from discoveries.

The way in which scientific ideas are conveyed is a natural part of the social history and geography of science. It is also of importance to the cognitive contents of science.[10] When science is conveyed from one milieu to another, it occurs by means of a selection process that decides which parts of science shall survive and which ones shall not. The scientist who understands how to market a new discovery is of no less importance than the discoverer. Stanislao Cannizaro (1826–1910) thus deserves just as much attention in the history of 19th century chemistry as does his more famous colleague Amedeo Avogadro (1776–1856). Avogadro proposed in 1811 the molecular hypothesis that today, in a somewhat altered version, bears his name. But Avogadro's suggestion was little regarded for half a century and only 'rediscovered' by Cannizaro in 1859. Cannizaro did not make any important discoveries himself but he marketed the molecular theory with vigour and considerable success. At a chemical congress in Karlsruhe in 1860 and at later occasions he argued the case of Avogadro's hypothesis which eventually became established as a cornerstone of chemistry.[11]

In some cases the salesmen, organizers and propagandists of science were not active scientists themselves. But even so, they have played a very important part in the development of science. An example of such a person can be found in Henry Oldenburg (1618?–1677), who was secretary and *primus motor* of the Royal Society in London. Oldenburg acted as a clearing house for the infant science of his day, collected and disseminated information throughout the whole of Europe at a time when periodicals had not yet become an established part of the system of communication in science. He was the founder of one of the first scientific periodicals, the *Philosophical Transactions of the Royal Society* (1665). Although Oldenburg never carried out any scientific research himself, he must be regarded as being as important a person as any in the history of science in the 17th century.

One way of organizing history of science is to divide it into 'horizontal' and 'vertical' sections (Figure 1). Horizontal history of science is understood here to mean the study of the development through time of a given, narrow topic; a scientific speciality, a problem area or an intellectual theme. In some cases it is possible to identify the origin (t_0) and the 'death' (t^+) of the topic in which cases the time boundaries are given. In other cases the upper boundary is the present day (t'). This case appears frequently since the reason for tracing a particular topic backwards in time is often tied up with the present importance of that particular topic. Horizontal history is typically discipline history or history of a sub-discipline.

Vertical history is an alternative way of organizing history of science materials. The vertically inclined historian starts out from a perspective that is more interdisciplinary in nature where the science that is in focus is seen as merely one element in the cultural and social life of a period. An element that cannot be isolated from other elements of the period and which, together with these, characterizes the 'spirit of the age' that constitutes the real field of this type of history of science. While horizontal history is a film of a narrow part of science, vertical history is a snapshot of the overall situation.

In horizontally organized history the historian isolates a particular discipline or problem from other, contemporary disciplines. This approach involves the danger of falling into anachronisms,

Figure 1. Two ways of organizing history of science. The topic along the vertical axis may be a scientific discipline, a problem area or a conceptual theme.

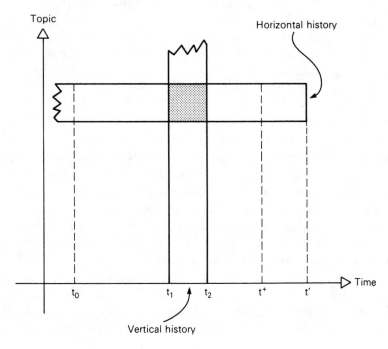

relying as it does on an assumption of disciplinary continuity. If the historian applies a narrow, horizontal perspective, the dependency on problems that lie outside the specialist subject may not be revealed. Disciplinary, horizontal history tends to become a bloodless recapitulation, a record of the origin, development and decay of the internal aspects of the discipline. As such it will not only be relatively uninteresting but also artificially confined. The historian of mathematics who studies the development of geometry, cannot allow himself to study pure geometry alone; he must be prepared to study the histories of art, architecture, philosophy, cartography, physics and perhaps several other fields.

Usually a scientific speciality of a certain period will be connected to, or have things in common with, other elements of the period. It is this complex whole, which Laudan calls a research tradition, that constitutes the actual units of history of science.[12] According to Wolf Lepenies,[13]

> It is not possible to write a history of a discipline without taking into account developments in neighbouring disciplines, whether they have been models or rivals for it. . . . the isolated history of a single discipline is completely repudiated by the historical study of science.

A similar attitude is expressed by David Knight:[14]

> To recover contemporary judgements, to see science as a continuing activity, and to make sense of the sources, one must study a wide spectrum of science in a relatively brief period. Historians have tended to drill a small hole down from the present through the strata of history; they should be well advised instead to look much more closely at the contents of one particular stratum.

In terms of modern divisions of science previous scientists often worked across disciplinary boundaries. They did not consider the boundaries between disciplines to be very sharp. For example, Copernicus should not be considered as only an astronomer; a label that would have astonished his contemporaries and, indeed, Copernicus himself. Copernicus was a canon in a cathedral chapter, he studied medicine and law, he occupied himself with theoretical and practical economics – and he was also interested in astronomy. If one isolates Copernicus the astronomer from Copernicus the official, the doctor, the lawyer and the humanist, one will not only give a distorted picture of the Polish scholar. One will also cut oneself off from possible vertical connections between the astronomical views of Copernicus and the activities that otherwise dominated his life. This also applies to most other scientists from earlier times: Buffon was not only a natural historian, Maupertuis not only a physicist, Herschel not only an astronomer, Steno not only a geologist, Priestley not only a chemist.

In spite of the criticism that can be raised against horizontally organized histories of disciplines, it would be wrong to follow Lepenies in repudiating this approach completely. At least in some cases it is possible to identify disciplines and specialist themes in earlier periods without committing sins of anachronism. It is just that these themes will only rarely be identical to modern themes and only rarely be unchanged throughout long periods of time. The risk one runs in cutting oneself off from important vertically integrated connections depends on the period and discipline being studied. An increasing disciplinary isolation is characteristic of the kind of highly organized, specialized science that has developed since the turn of the century. As far as modern science is concerned, it is, therefore, less problematic to organize history horizontally. Whether one needs to adopt a vertical, cross-disciplinary approach is not a matter of principle but of historical contingency.

While vertically organized historiography avoids the problems connected with identifying a stable discipline throughout a longer period of time, it lays itself open to other problems. The historian who follows the advice given by Knight and Lepenies and investigates the science of a short period of time, including its integration with intellectual life and society in general, will perhaps cut himself off from acquiring knowledge about the historical causes of the situation being analysed. The degree of arbitrariness in the choice of period or discipline complex will often be no less than the degree of arbitrariness to be found in the horizontally inclined historian's marking out of his field.

A special kind of organization of history that contains both horizontal and vertical traits is connected with the *thesis of invariant historical themes,* or the invariance thesis, for short. This is the thesis that history can be viewed as a variation on a relatively small number of constant themes or unit-ideas that manifest themselves at different times in all-important branches of culture. According to Arthur Lovejoy, who was an important spokesman for the invariance thesis in the history of ideas, unit-ideas can be compared with atoms of elements: just as the hundreds of thousands of chemical compounds can be understood to be combinations of a few kinds of atoms, the complex and extremely varied forms in the history of ideas can be conceived as combinations of a few unit-ideas.[15] Since it attempts to integrate different elements that make up culture and to simultaneously follow these through time, the thesis can be regarded as an attempt to circumvent the conflict between horizontal and vertical historiography. Lovejoy describes the thesis as follows:[16]

> The postulate . . . is that the working of a given conception, of an explicit or tacit presupposition, of a type of mental habit, or of a specific thesis or argument, needs, if its nature and its historic role are to be fully understood, to be traced connectedly through all the phases of men's reflective life in which those workings manifest themselves, or through as many of them as the historian's resources permit. It is inspired by the belief that there *is* a great deal more that is common to more than one of these provinces than is usually recognized, that the same idea often appears, sometimes considerably disguised, in the most diverse regions of the intellectual world.

Since Lovejoy, the thesis of invariant unit-ideas has been developed by many authors. One of these is Mendel Sachs, a physicist and philosopher. He writes:[17]

> It is my thesis that the actual truths sought by the philosopher and the scientist about the real world emerge in the form of abstract, invariant relations that are independent of the domain of understanding to which they may be applied, whether in the arts, the sciences, the philosophy of religion, or any other intellectual discipline, and that these relations are invariant with respect to the different periods of history during which they may be expressed. In the language of theoretical physics, I am contending that the *principle of relativity* – the assertion that the laws of nature are independent of the frame of reference in which they may be expressed – applies equally to the relations that govern the evolution of human understanding, i.e. the history of ideas, as it does to the natural phenomena of the inanimate world of the stars, planets and elementary particles.

According to Sachs, this thesis is supported by an examination of the theological and philosophical views of Maimonides (1135–1204) and Spinoza (1632–1677). These views reveal themselves to be 'analogous' or 'in agreement' with modern field theories in physics as they appear from Faraday to Einstein. The field concept is thus taken to be an invariant unit-idea. In the same way, several historians have fixed on what they consider to be striking similarities between concepts in classical natural philosophy and in modern science. Thus, Sambursky thinks that the concept of space in the theory of relativity is 'not unlike' that to be found in Aristotle; the *pneuma* of the Stoics bears a 'close resemblance' to Newton's ether and is 'not altogether unlike' the field concept of modern physics.[18] He concludes that 'the inner logic of scientific patterns of thought has remained unchanged by the passage of centuries and the coming and going of civilizations'.[19]

The thesis of invariance has been developed into a so-called thematic analysis by Gerald Holton.[20] According to Holton, one can profitably interpret pioneering scientific work as being based on underlying, possibly unconscious, concepts, methods and commitments that act as 'private' motives or restraints during the process of research. These *themata* are non-scientific in the sense that they are often not acknowledged by the scientist and rarely appear in official scientific discourse. The themata to which a scientist is committed do not necessarily stem from science. They can have been formed in early years or be the result of any sort of influence.[21] Like other forms of invariant ideas, themata do not have the status of theories. Their validity cannot be tested empirically or established by means of rational argumentation.

Holton's use of thematic analysis differs from the Lovejoy version of the invariance thesis in that it focuses on a short period of time and on individual scientists. In other words it is used vertically rather than horizontally. However, Holton believes that there are only a few themata in the history of science and that it is only very rarely that new themata arise. 'No matter how radical the advances will seem in the near future, they will with high probability still be fashioned chiefly in terms of currently used themata.'[22] The themata considered by Holton typically appear as opposing pairs of thesis–antithesis, such as evolution/devolution, plenum/vacuum, hierarchy/unity, reductionism/holism and symmetry/asymmetry.

However inspiring and interesting the invariance thesis is, it should be used with caution. Not as an infallible framework for organizing history but as a heuristic principle. In most cases it is problematic to talk of actually invariant unit-ideas as independent historical quantities. Unit-ideas are the result of comparative analyses fabricated by the historians. They are labels that imply that different works are analogous or belong to the same category. The selection of the historian and his interest in historical constancy may result in unit-ideas whose constancy in time is an illusion since the actual historical context in which they appear is disregarded.

Concepts and ideas are rarely or never quite the same over a long period of time. Although the names given to them by historians might be unchanged, fundamental concepts often develop beyond recognition through the historical process. Consider the doctrine of the divine nature of circular movement as developed in ancient Greece with special emphasis on astronomy. According to this doctrine, the planets have to move in circular orbits because the circle is the most perfect geometrical figure. The circle doctrine can reasonably be regarded as a unit-idea in the Lovejoy sense. It has played a dominant role in astronomy from Plato to Kepler and can be recognized in diverse cyclic models in and outside the natural sciences; in religious views, in economic theories, in physiology (Harvey's theory of the circulation of the blood) and in physics (Galileo's concept of circular inertia). But the idea of the circle as a natural, grand form of movement suddenly lost its magical appeal, at least in the natural sciences. This was a direct consequence of Kepler's discovery of the non-circular orbits of the planets. In modern science the circle doctrine has vanished. Thus, it is an example of a conceptual theme that has functioned for a long time as a unit-idea, but which is not really invariant. It is only in a very figurative sense that one can say that the circle doctrine has been of any importance in the science of the last 300 years. Modern cyclic models as those found in ecology and economics, for example, have nothing at all to do with the circle doctrine. Historians who attempt to trace the circle doctrine, viewed as an invariant idea, up to modern science are forced to interpret history of science in an artificial way.

The problem with using the invariance thesis over long periods of time is that it tends to press modern concepts and forms of thought down on earlier science instead of studying it in terms of its own premises. For example, the concept of continuity appears in both the stoic thinker Chrysippos (280–208 BC) and in 19th century mathematics and physics. With good will, it can be traced throughout the whole of the period that lies between. But the contexts in which the idea appears, and the meanings that are attached to the idea of continuity are certainly not the same in Chrysippos as they are in Gibbs or Cantor. There is no real historical information in asserting that the *same* unit-idea, in this case continuity, manifests itself in Chrysippos and in Gibbs. If it is the intention to say something about Chrysippos's thoughts as a historical phenomenon, then more or less forced analogies with much later occurrences do more harm than good.[23]

In a more vertical form of the invariance thesis as is to be found in Holton and others, there is less danger of anachronistic historiography. Here it is a question of locating particular themes in a particular scientist, and not of pointing out the temporal constancy of these themes. When an individual concerns himself with different things, with physics and economics, for example, it is only reasonable to suppose that the same general values and principles play a part in both the physical and economical activities, although they manifest themselves in different ways. It is therefore also reasonable to investigate the extent to which such principles can be identified in the various activities of the scientist; and whether a principle that arises in one activity might possibly have been transferred to another.

In an investigation of the prehistory of classical mechanics, Michael Wolff re-examines the impetus theory.[24] This theory, which was very influential in the late Middle Ages, states that a body in motion continues to move because of an impressed force or 'impetus' which is transferred from the mover to the moving body itself. While historians of science have traditionally considered the impetus theory to belong to physics, Wolff conceives it as a much more comprehensive idea that is equally at home in economic, technological and theological contexts. He therefore studies the impetus motif or the idea of transference causality as if it were an invariant unit-idea. The physical concept of impetus was closely tied up with economic considerations in the late-classical thinker Philoponos (around AD 500). The impetus motif can also be found again in Oresme (*c.* 1300–1385) and Buridan (*c.* 1320–1382) in both natural philosophy and in economic contexts. Although the impetus motif that appears in Oresme and Buridan is not the same one as Philoponos's, the very fact that Oresme, Buridan and others were not merely theologians and natural philosophers but also worked on economic problems is reason enough to ask whether the impetus theory is related to these philosophers' view on economics. In other words, whether there are isomorphic elements in the views of Oresme and others on natural philosophy and economics. Wolff argues that the principle of transference causality is such an isomorphic element and that, furthermore, the medieval, physical impetus theory derives from the idea of impetus prevailing in economics and technology. No matter whether Wolff's thesis is tenable or not, it is valuable to extend the perspective of history of science in such a way that fields like economics and technology are also taken into consideration. Scientific fields ought not to be studied in isolation or connected only with currents in the field of ideas as has been the tendency in the Lovejoy–Koyré tradition.

9

Anachronical and diachronical history of science

According to the *anachronical* view, the science of the past ought to be studied in the light of the knowledge that we have today, and with a view to understanding this later development, especially how it leads up to the present. It is considered legitimate, if not necessary, that the historian should 'intervene' in the past with the knowledge that he possesses by virtue of his placement later in time. Anachronical historiography, in the sense being used here, involves a certain type of anachronism; but it is not necessarily anachronistic in the usual, derogatory sense.

Today, anachronical history of science is rarely a conscious historiographical strategy. On the contrary, there is broad agreement about praising a non-anachronical ideal. Even so, in practice, anachronical history of science is widespread and difficult to avoid. The doctrine is connected to the presentist view of history which may be seen as a theoretical justification of anachronical historiography. Furthermore, this perspective is legitimate from the points of view that regard the goal of history of science as primarily bound up with the present situation (cf. Chapter 3). If one believes that it is the task of the historian of science to understand the technical contents of older science and to pass this understanding on to the scientists of today, then a way of presentation that is anachronical in tendency will be natural. A text will then be taken to have been understood if its true contents, in the current sense, can be represented with modern formalism and using modern knowledge.

Several studies of the history of thermodynamics have followed this prescription.[1] Truesdell has made a virtue out of this kind of anachronical historiography where the focus is on a logically satisfying reconstruction of how earlier science could have looked. In the opinion of Truesdell, modern knowledge of thermodynamics is a precondition for writing its history:[2]

> Only now do we know a decent theory of the scope the creators
> sought, so only now can we see just where the old authors stopped
> short or even went wrong. . . . much of what I write now about
> the classical papers on thermodynamics I could not have written
> twenty years ago, because I did not then have the grasp of rational
> thermodynamics that today we may and do teach our beginning
> students. This knowledge does not change the historical record one
> whit; rather, it teaches us to read it better.

Truesdell's view of the history of science is not shared by many
of today's practitioners of the subject. If one considers Truesdell's
aim, however, that of writing history of science for scientists and
as clarification of concepts, it is, at least, a consistent attitude.

The *diachronical* ideal is to study the science of the past in the
light of the situation and the views that actually existed in the
past; in other words to disregard all later occurrences that could
not have had any influence on the period in question. Occurrences
that took place before, but which were actually unknown at the
time, have to be regarded as non-existent too.

So, ideally, in the diachronical perspective one imagines oneself
to be an observer *in* the past, not just *of* the past. This fictitious
journey backwards in time has the result that the memory of the
historian–observer is cleansed of all knowledge that comes from
later periods. The diachronical historian is therefore not interested
in evaluating the extent to which historical agents behaved rationally or whether they produced true knowledge in an absolute or
modern sense. The only thing that matters is how far the actions
of the agent were judged to be rational and true by the agent's
own time. In this sense, one may say that there is a relativistic
element in diachronical historiography. In many ways, Collingwood's view of history is in accordance with the diachronical
ideal, as appears, for example, in the following quotation:[3]

> History . . . meant getting inside other people's heads, looking at
> their situation through their eyes, and thinking for yourself whether
> the way in which they tackled it was the right way. Unless you can
> see the battle through the eyes of a man brought up in sailing ships
> armed with broadsides of short-range muzzle-loading guns, you are
> not even a beginner in naval history, you are right outside it.

How should the successes of past science be evaluated in relation
to its failures? Strictly speaking, this question is only a relevant
one within an anachronical perspective, since evaluations of succes-

ses or failures are evaluations of the extent to which theories put forward at a particular time are still regarded as valid or at least having once had a positive importance for contemporary views. Such evaluations as these lie outside the diachronical perspective where the present, in a manner of speaking, does not exist. On the other hand, evaluations of the success or failure of occurrences at the time are relevant in diachronical, but not in anachronical historiography. In the latter case, Kepler's laws would be seen as successful and pioneering however one looks at them, and the fact that they were not thought of in this way for the first 50 years of their existence will be considered quite irrelevant. In general, the result is a different selection and allocation of priority to the events of the past.

The anachronically inclined historian, when dealing with William Harvey's famous discovery of the circulation of the blood (1628), will legitimate it by stating that, though it had certain speculative characteristics, Harvey's theory has proved to be an essentially correct description of the passage of the blood in the body. Since the theory was the first version of the true explanation (i.e. the one accepted today), it will be judged a success and an important milestone in the history of medicine. The diachronically inclined historian, dealing with the same subject and trying to put himself into the situation of a person working about the year 1640, will be more cautious in his evaluation of Harvey's discovery. In fact, Harvey was ridiculed at first and his theory of circulation was met with much opposition and scepticism during the first decades. The historian will be interested in how Harvey's work was received at the time, for example, in the criticisms directed at the theory by Gassendi and others. And he will draw attention to the support given to Harvey by mystics and alchemists (such as Robert Fludd and Elias Ashmole) on a definitely unscientific basis. While Fludd will appear in diachronical historiography as a key person in connection with Harvey, he might not be mentioned at all in anachronical historiography.[4]

In anachronical historiography the subject matter of history of science is the same as the subject matter of science. Scientific facts and theories are regarded as having a permanent, almost transcendental existence even in periods when they were not recognized. In the words of Gerd Buchdal, anachronical historiography is based on 'the misleading presupposition that "science" (as against *scien-*

tia) is a quasi-object latently existing in all ages, signs or symptoms of which may be discerned to appear during any stages of world history'.[5] Accordingly, science becomes a phenomenon that is bound to make progress in the direction of truth. It is then the task of the historian to elucidate this development towards true knowledge as it takes place through successive experiments and theories. The philosophy of science that lies behind anachronical historiography leads to the temptation to write history backwards, to teleological history of science.[6] This is an approach that has been badly shaken by the criticisms put forward by Kuhn and other post-positivistic philosophers of science.

The teleological writing of history is given much attention by the French philosopher Gaston Bachelard (1884–1962) and by the thinkers inspired by him, including M.Fichant, M.Pécheux, D.Lecourt and G.Canguilhem. Like other modern philosophers Bachelard has strongly criticized what is called *historicism*, i.e. the view that the present is merely a result of the past of this present, a temporary terminus in a linear continuous development. However, a topical history of science, i.e. a history that is directly linked to the current level of science, is of great importance to Bachelard, to whom an 'antiquarian' interest in the past of science for its own sake is not real history of science at all. What has to be done is to replace the philosophically suspect historicism with another idea that will still ensure that the history of science retains its topical interest. Taking as given that the job of the historian of science is that of evaluating the value and truth of his subject, Bachelard writes:[7]

> In order to evaluate the past properly the historian of science must know the present; to the best of his ability he must learn the science whose history he plans to write. And it is through this that, whether one likes it or not, the history of the sciences has a strong connection with the science of the moment. It is when the historian of science is initiated into the modernity of science that he is also able to uncover more, and more subtle, nuances in the historicity of science. Consciousness of modernity and consciousness of historicity are strictly proportional here.

According to Bachelard, one has to double factual history with an evaluatory history where the criterion of value lies in the values of modern science. Bachelard proposed the term *recurrent history* for the non-teleological, active reflection of the science of the past in the light of contemporary science. Recurrence is an 'assimilation of past science through the modernity of science' that at the same time has the consequence that history is constantly rewritten.[8] This recurrent history is deliberatedly anachronical, since it decides whether earlier science is valid or not in the light of present knowledge; but it is not a continuist, teleological history.

Recurrent historiography is aimed at what Bachelard calls 'sanctioned history' (*histoire sanctionée*) that is seen as a double of traditional 'obsolete' history (*histoire perimée*), which merely describes earlier occurrences. According to Fichant, obsolete history is 'the history of the thoughts that have become unthinkable in present-day rationality', while sanctioned history is 'the history of the thoughts that are always topical or can be made topical if they are evaluated in terms of the science of the day'.[9] For example, Bachelard rejected the optical theories of Descartes as worthy objects of study for recurrent historiography because they are today regarded as false; on the other hand, the wave theories of Huygens and Fresnel belong under the sanctioned history because they have a permanent value as parts of the 'topical-past science'.[10]

Bachelard was aware of the fact that the use of the idea of recurrence can easily be overdone if it is not combined with 'real tact'. He believed that the recurrent perspective and the division into an obsolete and sanctioned history of science is, in the main, only justified in the later phases of a scientific development, the phase of modernity in which it has achieved relative autonomy and built up a corpus of criteria for evaluation. But in spite of these reservations, Bachelard and his school maintain that it is unavoidably necessary to evaluate history of science recurrently, since otherwise it will degenerate into a merely antiquarian history, of no relevance to the present. 'The historian of science is necessarily a historiographer of truth.'[11] The truth that he seeks is not the truth *about* history but the truth *in* history.

What we here call anachronical history is largely the same as what is known as the *Whig interpretation* of history. By this is meant 'the study of the past with one eye, so to speak, upon the present', according to Herbert Butterfield (1900–1979), who invented the term and identified it with 'unhistorical history writing'.[12] Butterfield's critique was originally aimed at a strong tradition in English political historiography in which the history of England was described as an unbroken progress towards the democratic ideals that the Whig party was said to represent. But Whig historiography soon passed into general use as a term (usually with negative overtones), and it has also been much discussed in history of science. In a paper which Butterfield wrote in 1950, the most important anti-Whig morals are drawn out as follows:[13]

> The whole fabric of the history of science is on the one hand lifeless, and on the other hand distorted if it is based on the principle of seizing upon one writer in the fifteenth century (Nicholas of Cusa for example) choosing him because he had some single idea that strikes us as wonderfully modern; then seizing upon another man in the sixteenth century (Leonardo da Vinci, shall we say), because he had a curious hunch or anticipation of what was to be produced by scientific research at a much later period. In reality I believe it has proved almost more useful sometimes to learn something of the misfires and the mistaken hypotheses of early scientists, to examine the particular intellectual hurdles that seemed insurmountable at a given period, and even to pursue these courses of scientific development which led into a blind alley, but which still had their effect on the progress of science in general. What is wrong in the history of science as in all other forms of history is to keep the present day always before one's mind as the basis of reference; or to imagine that the place of a seventeenth-century scientist in world history will depend on the question how near he happened to come to the discovery of oxygen.

We shall now discuss some of the ways in which anachronical Whig historiography may typically result in objectionable history of science.

(A) Evaluation and granting of status

If modern science functions as a mark book for earlier science, one will tend to present occurrences that can be seen today as pioneering as though they were just as pioneering in their historical situation. And one will evaluate the knowledge of the past as though it concerned the same subject and concepts that we think it was 'really' about today. We have seen how the retrospective evaluation of Harvey's idea of the circulation of the blood does not reflect historical reality at the time of the discovery. Another example is the following.

According to an idea that was popular among alchemists of the Middle Ages, all metals consisted of two principles, often called 'sulphur' and 'mercury'. In many alchemical writings one can read about the synthesis of various metals based on processes involving 'sulphur' and 'mercury' in suitable ratios. If one were now to make the rather obvious mistake of believing that the sulphur–mercury theory of the alchemists was a theory based on what we understand today by the elements of the same names, the theory will seem speculative and completely silly. The eminent chemist and historian of chemistry E.Meyer apparently made this anachronistic mistake since he wrote about the sulphur–mercury theory:[14]

> It is surprising that the chemists of the 13th and 14th centuries, whose knowledge of chemistry was rather comprehensive, were satisfied with this kind of speculation about the composition of metals without seriously attempting to represent the substances absorbed into these and other bodies.

If the sulphur–mercury theory is studied diachronically one will quickly realize that the 'philosophical sulphur' and 'philosophical mercury' of the alchemists have to be interpreted as principles or abstract ideas and not as material substances. Given the meaning that the alchemists themselves placed on their 'sulphur' and 'mercury' their theory was far from being foolish; it was a rational idea supported by experiments. In a diachronical perspective, instead of being a fantastic speculation it becomes quite a reasonable theory.

(B) Formalization

Just as historical materials can be modernized by means of translations of words and concepts, modernization can also take place in the form of a formalization – usually mathematical – of statements that were originally expressed in a non-mathematical form; or in a mathematical form that was different from ours. There need not be anything unhistorical in either modernized translations or in conversions into mathematical forms as long as the conceptual contents are not significantly changed from those of the original. It is, after all, the task of the historian of science to transform and communicate older science to a present-day public, which means that it can be necessary to formulate historical statements in modern terms in order to make the past at all understandable. Modernization can, however, easily result in serious anachronisms that distort historical reality beyond recognition.

As an example, let us have a look at what is sometimes called Aristotle's law of motion. According to Aristotle, a body moves because it is influenced by a motive force (F). The speed (v) is proportional to the force and inversely proportional to the friction (R) between the body and the medium in which it moves. Thus, Aristotle's law of motion can supposedly be expressed by the equation[15]

$$v = k \cdot F/R,$$

where k is a constant. However, this is an anachronism on three levels. In the first place, the mathematical form was alien to Aristotle and his time. Not just the form, but the very idea that motion can be expressed quantitatively, was outside the framework of Aristotelian science. Secondly, the terms contained in the law ('force', 'speed', 'friction') refer to knowledge and concepts that only came into being much later. And thirdly, the status attached to Aristotle's ideas about moving bodies is unhistorical. The concept of natural law, in the sense it is found in Newton, for example, did not exist at all at the time of the ancient Greeks. Unless one attempts to view the ideas of Aristotle in a diachronical perspective one will be tempted to compare the value of Aristotle's (fictitious) law of motion with those of Galileo or Newton. This is clearly unreasonable.

The same type of problem has been discussed by Cohen in connection with Newton's second law of motion.[16] This law is usually stated as $F=m \cdot a$, where m is the mass of the body and a its acceleration. Newton never formulated his famous law as $F=m \cdot a$ or in any other way that reminds one of the later, institutionalized textbook version. Furthermore, Newton used the word 'force' in such a way that it probably ought to be translated into modern English by the word 'momentum' rather than 'force'. If one projects the modernized version of the second law down on Newton, his own version will seem non-understandable:[17]

> The change of motion is proportional to the motive force impressed; and is made in the direction of the right line in which that force is impressed. . . . If any force generates a motion, a double force will generate double the motion, a triple force triple the motion, whether that force be impressed altogether and at once, or gradually and successively.

Cohen's point is not that there is in fact a disagreement between Newton's authentic law and the modernized version. But rather that the apparent disagreement can only be understood in a diachronical perspective. He writes: 'The historian's job is rather [in contrast with that of the philosopher] to immerse himself in the writings of scientists of previous ages, to immerse himself so totally that he becomes familiar with the atmosphere and problems of that past age.'[18]

(C) Coherence and rationality

It will normally be reasonable to suppose that the thoughts of a scientist, as they appear in publications, are coherent and consistent. But what if the historian comes across texts that are apparently marked by an absence of coherent and rational thought? In this situation, which occurs quite frequently, the lack of coherence can be evaluated in one of three ways:

(1) The lack of coherence is accepted as a valid expression of the fact that the thoughts of the agent were not, in fact, coherent; that they were unsystematic, confused and perhaps inconsistent.

(2) The lack of coherence is thought to be merely deceptive. The account is assumed to be coherent in reality and it is then the task of the historian to interpret the text in such a way that the coherence can clearly be seen. When this kind of interpretation has been made, the actual thoughts of the agent are understood.

(3) An attempt is made to resolve the lack of coherence by studying the event in more detail and by placing it in its proper historical framework. Unlike in (2) one will not force out the coherence, but merely assume coherence as a reasonable working hypothesis. If the event still seems to be incoherent after further investigations one will assume that the position in (1) is the most reasonable one.

As Skinner has pointed out, the attitude expressed in (2) easily leads to motives and thoughts being assigned to the historical agent for which there is no documentary evidence and for which the historian is really responsible.[19] 'The mythology of coherence', as Skinner calls it, results in explaining away instead of explanation. Why not accept that scientists do not *always* argue in a way that is clear, coherent and consistent? Many discussions about how far a scientist really thought the one thing or the other are futile precisely because they are based on the postulate that he must have meant the one thing *or* the other; instead of recognizing that the scientist might have had conflicting views of the same matter or have meant the one thing in one context and the other thing in the other context. That this might have been the case is not to deny the scientist rationality; he might have had good grounds for meaning both.

Skinner's attack on the mythology of coherence is based on a purist diachronical view of history according to which past texts are solely about the past and are completely without modern relevance. What matters, according to Skinner, is not the validity or present significance of ideas but only the *intention* of the author, viewed in the historical context. The historian should focus on what the author consciously wished to communicate to his readers and, therefore, in order to understand a text he should grasp its subjective intention. I shall discuss Skinner's theses separately. First his contextual purism.[20]

The emphasis on the intentions of authors and the corresponding neglect of their impact are likely to produce a fragmentary picture of the history of ideas, reducing it to a series of isolated events. In such a picture there is no room for real historical inquiry. It is a fact, well known from common experience, that the actual implications of thoughts do not always coincide with the intentions behind them. Authors may simply fail to grasp the potentialities of their own thoughts. Even if a thinker had no intention of stating a certain doctrine, he may well have done so in fact if contemporary or later thinkers understood him to have done so. This is a historical fact of much more significance than the author's subjective intentions.

An example which shows that there is more to history than what is implied by the intentions of the agents is supplied by the British natural philosophers of the 17th century. The group of *virtuosi*, including Boyle, Wilkins, Ray, Barrow and Newton, were Christian philosophers who firmly believed that the new science was a defence against materialism and atheism. They all intended to strengthen Christianity and denied that there could possibly be a conflict between Christian belief and science.[21] Should the historian rest content with reporting these intentions? If so, he would surely be unable to explain the later development and would miss an opportunity to understand the natural religion of the 18th century, for example. Despite their intentions, the actual effect of the mechanical natural philosophy was to weaken Christianity and lay it bare to attacks from materialists.

As to the mythology of coherence, Skinner's view cannot easily be dismissed. Undoubtedly much history of science commits anachronistic sins by streamlining and clarifying past thoughts far beyond what is justified by textual evidence. Even so, it would be problematic to accept Skinner's view without qualification. Scientists are not fools, nor were they in the past. If one discovers foolishness, obvious absurdities or inconsistencies in the analysis of a text, one ought not immediately to accept it as an expression of the true historical character. On the contrary, one ought to be suspicious and have doubts as to whether one has properly understood the text. It could well be that the absurdities and inconsistencies are the result of an anachronical reading and that they do not exist in the proper diachronical perspective.

Attempts to rationalize and modernize earlier events are often bound up with the thesis of coherence and also with the doctrine of anticipation mentioned below. In what are now scientific fringe areas like astrology and alchemy, especially when such areas have also been cultivated by great scientists, it is tempting to rationalize and to view these areas as almost modern scientific theories that were just expressed in an odd way. If a view like this is adopted it becomes the task of the historian of science to draw out the rational kernel that is assumed to be there.

Some analysts of Newton's alchemical works have argued that they 'really' contain an atomic theory that resembles the modern one. Thus Karin Figala thinks that Newton's ideas 'show ... surprising resemblances with the atomic shell model proposed by N.Bohr'.[22] From her rationalization of Newton's alchemy, Figala even thinks that Newton 'almost seems to have suspected' that there are metals that are more precious and have greater density than gold. Platinum, iridium and osmium are thus thought to lie indirectly in Newton's scheme although these elements were only discovered decades after Newton's death. Figala's rationalization of Newton is a clearcut example of anachronical history of science. Newton did not, of course, have any idea at all of the existence of planet-like, compound atoms or of new elements in any modern sense of the word. The ingenious reconstructions of Figala would have been totally incomprehensible to Newton and his age.

While the assumption of coherence often involves anachronic elements, it might just as well turn up in connection with diachronic historiography. In fact, considered as a criterion of progress in historical reconstruction it is rather part of the Koyréan contextualist tradition, which is definitely based on diachronical virtues. In this tradition, clarity and coherence of scientific thinking are often taken for granted.[23] Kuhn, for example, who owes much to Koyré, openly advocates the assumption of coherence. In a work on the genesis of quantum theory, Kuhn concluded that Max Planck's celebrated theory of 1900 did not really constitute a revolution in physics. In defence of his reinterpretation of Planck's theory he argues that[24]

> the reinterpretation makes the development of Planck's black-body research both more nearly continuous and also a deeper, more elegant piece of physics than it appears in the standard version . . . the Planck who appears in the reinterpretation is a better physicist – less a sleepwalker, deeper and more coherent – than the Planck of the standard story.

In defence of the diachronically based assumption of coherence, Kuhn has pointed out that much of the criticism rests on a doubtful notion of what scientific discoveries are. If the scientist is said to have had only a confused view, that he came to his result stumbling like a sleepwalker, the process of discovery is accounted for by judging it from the subsequent formulation, not yet worked out.[25] From a historical point of view this notion of discovery makes no sense.

A distinction should be made between the *assumption* of coherence and the *dogma* of coherence. While the latter invariably leads to bad historiography, the former can be a fruitful strategy when properly used. One should furthermore distinguish between the anachronical case, in which clarity is judged by modern standards, and the diachronical case, in which the assumption is defended without use being made of hindsight. Surely the historian is entitled to clarify obscure passages if he can justify the clarification by means of independent evidence. But he should not rule out the possibility that the text might, in fact, be obscure.

(D) Anticipation

There is a long tradition in the history of science of taking an interest in which persons or theories were the forerunners of a particular later theory. This interest has recently been criticized by many authors.[26] But the criticism is by no means new. It was formulated very precisely by the French physicist Jean-Baptiste Biot (1774–1862) 150 years ago:[27]

When an important and influential new event, whose certainty is assured and whose range is proved by its use, appears in the scientific world, then the contemporaries, due to a natural habit, tend to explore with curiosity whether there are traces of it to be found in the past. If they find some, even if they are uncertain, they reach out for them and revive them with a mixture of gullibility and hindsight. This critical work is of great merit when it is just: justice should be done to unrecognized inventors. However, it is also necessary to adopt their point of view and to understand the expressions they used, in the same way as their own age understood them; their ideas should be given the same range that they gave them, and finally the unchangeable rules of scientific discussion must be applied to their results. So one must distinguish carefully between assertions and proofs, between individual cases and established truths; for it would be neither useful nor just nor philosophical to accept as proved in an older writer what one will not allow as hypothetical in a contemporary writer.

The point being made by Biot is partly that assertions about anticipation necessarily involve speculative interpretations directed by later knowledge. And partly that scientific discoveries ought to be judged with respect to their actual historical significance: discoveries can only be regarded as effective if they have achieved a widespread acceptance. Notice, however, that Biot puts forward the non-diachronical view that earlier science ought to be judged according to the same criteria as modern science, viz. 'the unchangeable rules of scientific discussion'.

By its very nature, the idea of anticipation involves an anachronical perspective. In itself this may not be problematic; but it becomes so if clairvoyant abilities are ascribed to predecessors and if later theories are projected on to the works of predecessors. If these pitfalls are not avoided, the result is pure anachronism. As happens when the French scientist Pierre Maupertuis (1698–1759) is presented as the forerunner of all the biology that was developed over a hundred years later:[28]

> ... Maupertuis was most clearly gifted with prevision ... virtually every idea of the Mendelian mechanism of heredity and the classical Darwinian reasoning from natural selection and geographical isolation is here combined, together with De Vries' theory of mutations as the origin of species, in a synthesis of such genius that it is not surprising that no contemporary of its author had a true appreciation of it.

The problem about the concept of anticipation is, of course, that to a high degree it is the historian's interpretation of the forerunner that decides to what extent there is a connection between the alleged forerunner and the later doctrine. This is an unavoidable element in anticipation historiography. There is no single criterion for when P is said to have anticipated N apart from the obvious fact that P and N, in some way or other, must be concerned with the same subject. According to Sandler, the following cases can be singled out:

I P was in the same disciplinary tradition as N and actually influenced N; but P formulated the doctrine in a way that was incomplete and that did not win immediate recognition. This form is usually unproblematic, but lies on the borderline of proper anticipation. Instead of saying that P anticipated N, one can say that P influenced N or that N developed further the thoughts of P.

II But P can also be placed in quite a different disciplinary tradition to that of N and still be called a forerunner of it. Thus, in 1798, Thomas Malthus suggested that human population would invariably outrun the food supply if human numbers were not kept down by political or moral means. Malthus's theory was concerned with politics and economy, not biology. Nevertheless, Malthus is often placed in a biological context and called a forerunner of parts of Darwin's theory, whose creation was, in fact, inspired by Darwin's reading of Malthus.

III P need not have any inkling of N's doctrine, and can have been opposed to the way of thinking it expresses. P can have been a forerunner against his will. This was the fate of Boscovich (1711–1787) in relation to the theory of matter which Priestley worked out in the 1770s. In Priestley's theory matter and spirit were not two distinct kinds of substances but reducible to the same 'powers' and thus basically identical. Boscovich's theory, which served as an inspiration to Priestley, was not materialistic in Priestley's sense and Boscovich in fact protested against the use Priestley made of it.[29]

IV P need not have had any influence on N, neither direct nor indirect. N need not have had any awareness of the existence of P. In this case it is only the historian who claims that P is the forerunner of N, arguing that there is an objective connection between the subjects of P and N. We can find an example of this in Servetus (*c.* 1511–1553), who wrote a theological work in which the circulation of the blood is discussed in opposition to the doctrine of Galen. On the basis of this Servetus is often referred to as a forerunner of Harvey. But Servetus and his books were burnt at the stake for heresy and his thoughts were therefore unknown to those who followed him, including Harvey (today, only three copies of Servetus' book are known to exist).

V Finally, the word 'anticipation' is sometimes used in the sense of 'prediction'. This is unfortunate since prediction is a different, more precise term. If P predicted N the relationship is one between theory and discovery, not a relationship between different formulations of the same doctrine.

As Sandler has pointed out, anticipation is a context-dependent concept that will often be evaluated differently by scientists and historians. Of the types mentioned above, II, III and IV will rarely be accepted by scientists as anticipations. When the historian shows interest in such cases, it is because they may give rise to interesting questions. In this case they are heuristically valuable. Why did P remain unknown to N? (case IV); in what connection did N know about P and how was P's disciplinary tradition carried over into N's? (case II); why did P oppose the formulation of his idea received in N? (case III).

Anticipation historiography is closely connected with the thesis of invariance and, in general, with continuity history of science. If scientific development is seen as a continuous, conservative process then the search for direct predecessors becomes a central task for the historian. This method, in which a development is presented as a sequence of small changes and in which it does not, therefore, have any clear beginning, has been called the emergence technique.[30] It is found in pure form in Duhem:[31]

> History shows us that no physical theory has ever been created out of a whole cloth. The formation of any physical theory has always proceeded by a series of retouchings which from almost formless first sketches have gradually led the system to more finished states . . . . A physical theory is not the sudden product of a creation; it is the slow and progressive result of an evolution.

Therefore, when Duhem is to tell the history of the theory of gravitation, he describes it as an unbroken chain of predecessors to Newton, starting with the ancient Greeks. Among the many predecessors included in Duhem's emergence chain are examples of all the types mentioned above.

Can one conclude from the misery of Whig historiography that all anachronical elements ought to be avoided and that history of science ought to be dealt with from a purely diachronical perspective? As has already been indicated, the answer is no. A totally diachronical history of science would not be able to live up to the demands that are normally made on historical expositions. It might perhaps give a true representation of the past but it would also be antiquarian and inaccessible to all but a few specialists. Several authors, in fact, have warned against carrying anti-Whig historiography to extremes. In Merton's view, the time has come for an anti-anti-Whig revaluation.[32]

Diachronical historiography can only be an ideal. The historian cannot liberate himself from his own age and cannot completely avoid the use of contemporary standards. During the preliminary study of a specific period one cannot use the period's own standards for evaluation and selection; for these standards form part of a period that has not yet been studied and they will only gradually be revealed. In order to have any kind of view at all of one's subject one has to wear glasses; and these glasses must, unavoidably, be the glasses of the present. The historian cannot rely purely on criteria of significance accepted in the past. Only in a few cases will there be an undisputed consensus on priority in the past; usually the establishment of consensus will involve selection and hence imply the historian's intervention.

In many cases it will be the obvious thing to do to use modern knowledge in the analysis of a historical event. By so doing one may be led to interesting questions that could not be formulated on a purely diachronical basis. Thus, in the opinion of most historians, it is interesting to ask why the Greeks did not discover the irrational numbers (such as $\sqrt{2}$), a problem central to the understanding of the foundation crisis of Greek mathematics. But a question like this can obviously only be put by somebody who knows that rational numbers can be extended with irrational numbers in the way that happened much later. It is a question that could not have been asked by the Greek mathematicians. In general, why-not questions have no place within strict diachronical historiography.

Similarly, it is only in retrospect that many important connections manifest themselves. Around the year 1845 several scientists (Mayer, Colding, Joule, Helmholtz) formulated doctrines about what was *later* known as the constancy of energy. But at the time it was not at all clear that the discoveries were 'really' about the constancy of energy or that they were about the same phenomenon at all. A historian who places himself mentally in the year 1847 will not be able to see the connection between the discoveries of Mayer, Colding, Joule and Helmholtz, and will therefore be unable to treat these discoveries collectively. It is only if one allows an anachronical perspective that it can be seen that they were, in fact, a case of the 'same' discovery.[33] This perspective can, of course, easily ruin historical understanding if, when interpreting Mayer for example, one has at the back of one's mind the whole time that his work was 'really' about energy conservation.

Extreme diachronical history will clash with the pedagogical dimension that forms an integral part of all historical research. The history of science is not a two-part relationship between the historian and the past, but a three-part relationship between the past, the historian and a present-day public. On the whole, diachronical historiography will fail to perform its function of communication. It will have a tendency towards merely being a detailed but passive description of historical data, while analysis and explanation are neglected. This tendency can be found in Skinner and in Butterfield, the latter making himself a spokesman for narrative explanation:[34]

> In the last resort the historian's explanation of what happened is not a piece of general reasoning at all. He explains the French Revolution by discovering exactly what it was that occurred; and if at any point we need further elucidation all that he can do is to take us into greater detail, and make us see in still more definite concreteness what really did take place.

The historian of biology, D.Hull, has pointed out that distortions cannot be automatically avoided by 'forgetting the present' or by pretending that present knowledge does not exist.[35] Instead of carrying out this play-acting the historian ought to admit that in many cases he has knowledge of the received judgements of history and openly use this admission to prevent proper anachronisms and at the same time render his studies understandable and of interest to a modern public.

The objections that can be raised against *strict* diachronism does not imply that the historian is forced to look at the past with modern science as his starting point. Neither should they be taken as support of relativist or presentist historiography in its extreme form. At least to some extent, the diachronical perspective is able to supply history with a measure of objectivity that does not depend on time or fashions. As a methodological guide and an antidote to the pitfalls of Whig history, the diachronical ideal is indispensable.

Histories of the same subject may be radically different according to whether they are written from an anachronical or a diachronical point of view. Thus, since the beginning of this century Gregor Mendel (1822–1884) has been judged to be an unappreciated pioneer in the history of genetics. The true nature of his contributions, the Mendelian laws, was not properly understood at the time when the laws were formulated, but only since the so-called rediscovery in 1900. This rediscovery of Mendel not only meant that in one leap he advanced from being an obscure minor character to being a leading actor in the history of biology; it also implied that the evaluation of what Mendel had actually performed in 1865 changed. The history of biology was rewritten in the years after 1900. Not because there were any new source materials available about Mendel, but because he was now seen in a new perspective. It is only in the light of later developments that there is any sense in asserting that Mendel was unappreciated by his contemporaries. If we attempt to read the works of Mendel strictly in their diachronical context, they will appear to be rather orthodox contributions to the plant improvement tradition in botanical research, and not as a revolutionary anticipation of genetics. True, many of Mendel's experiments and interpretations were novel and he felt himself that their originality was not recognized by his contemporaries. But since nobody else shared Mendel's view of his own work, its originality does not belong to diachronical history. In a diachronical context Mendel was not misunderstood by his age, but understood.[36]

It appears, then, that one has to operate with two Mendels. The one seen in the perspective of his own time; the question 'why was Mendel ignored or misunderstood by his own time?' will not apply here. The other Mendel is the Mendel of the 20th century, the originator of the genetic laws. It is in this context that the question will be asked. But it is a question that in reality should be read 'why was it believed after 1900 that Mendel had been ignored or misunderstood?' In this version it is just as much concerned with the knowledge of genetics at the beginning of the 20th century as with the historical Mendel.

We conclude that in practice the historian is not confronted with a choice between a diachronical *or* an anachronical perspective. Usually both elements should be present, their relative weights depending on the particular subject being investigated and the purpose of the investigation. The historian of science has to be a person with the head of a Janus who, at the same time, is able to respect the conflicting diachronical and anachronical points of view. According to the Dutch historian of science, Hooykaas:[37]

> In order to judge fairly, the historian has to approach the thinking, observing and experimenting of the forebears with a sympathetic understanding: he must possess a power of imagination sufficiently great to 'forget' what became known *after* the period he is studying. At the same time, he must be able to confront earlier views with the actual ones, in order to be understood by the modern reader and in order to make history something really alive, of a more than purely antiquarian interest.

10

Ideology and myths in the history of science

Histories of science involve particular perspectives, aims and methods of organizing materials that do not arise out of the objectively given past itself. Very often, history of science also serves a legitimating function. The fact that histories are written with commitment and from a particular motive, or may serve legitimating functions, does not necessarily imply that they are products of bad historiography (see also chapter 5). But as soon as documentary evidence is distorted, ignored or allocated disproportionate importance in order to fit in better with a particular moral that serves a social function, history becomes ideological.

I shall use the term 'ideology' in the sense that an ideological doctrine is a doctrine which legitimates the views and interests of a particular social group. This is a necessary but not a sufficient condition. The doctrine must also give a distorted or misrepresented picture of the reality it refers to. According to Althusser, an ideology is 'a statement which, while it is a symptom of a reality that is separate from the reality it refers to, is a false statement in so far as it touches on the object it has in view'.[1] The bias that is connected with ideological doctrines can be deliberate; but it will not normally be so. Ideologies are rarely admitted by the ideologists, nor by the social group to whose interests the ideology is directed.

Ideological historical writing covers a wide spectrum.[2] At the one extreme there are outright ideological histories which serve, for example, political purposes. These 'external' ideologies are directed to the lay public or political bodies, serving a wider political function. They may legitimate particular political systems by representing them as superior with regard to scientific development; or they may legitimate science by means of arguments of utility or cultural value. This is one of the ways in which history of science may play a role in science policy (see also chapter 3).

The 'external' ideologies should be distinguished from the 'internal' ideologies which are primarily directed to the scientific community. Internal ideologies serve legitimating functions too, but usually in a subtler, less 'political' way. More appropriately, perhaps, one should speak about mythicization in history of science. Myths are socially useful doctrines that are only indirectly related to historical fact. The social function of the myth typically lies in its strengthening of the prestige, unity and self-consciousness of a social group, in this case the practitioners of a scientific discipline. An event is made into a myth when it is ripped out of its actual context and is given a meaning that makes its social function possible. While myths and ideologies often legitimate status quo, they may also serve progressive functions. As we shall exemplify below, histories of science may be written to prevent change as well as to justify change. In themselves, myths are neither conservative nor progressive.

'External' ideological history of science is typically to be found in connection with nationalistic and ethnocentric historical writings. There is a long tradition behind such writing. That it exists ought not to be a cause for surprise: history of science is no less sensitive to cultural and political crises than so many other intellectual institutions. History of science is just one of many instruments that a people or a nation can mobilize in a time of crisis for the waging of ideological propaganda warfare.

During and immediately after the First World War, the hostility between the belligerent parties resulted in histories of science that were markedly nationalistic. For example, the eminent French physicist and mathematician Émile Picard wrote a history of science in 1916 that was to show that all that was good in the development of science was due to French scientists and all that was bad to German scientists.[3] Another physicist, the Nobel Prize winner Philippe Lenard (1862–1947) wrote an 'Aryan history of science' 20 years later, based on the Aryan or *völkische* view of science that Nazi Germany wanted to develop.[4] Lenard's historiography manifested itself in, among other things, attempts to distinguish between so-called Aryan and so-called Jewish contributions to science. Lenard argued that all the positive contributions to the history of science had been made by Aryans, while the many great Jewish scientists had either carried out bad research or had stolen their good ideas from non-Jews.

Nationalistic mythicization received a special stamp in the Soviet Union from about 1930 to 1955. History of science was used ideologically, as a defence of the political system and in order to increase Russian national pride. It was hoped to help counteract the Soviet feeling of cultural and scientific backwardness by the use of a history of science designed for the purpose. This history was marked by, among other things, xenophobia and the assertion of a series of priority claims.[5] The close connection with scientific progress that Soviet communism claimed for itself had to be expressed in a history that legitimated this connection, in other words showed how scientific progress in Russia had only arrived with communism. Soviet history of science of the period was partly a communist history but it was also a national Russian history that took up a hostile attitude to what was officially regarded as the West's unfair dominance in the field of science. As in Nazi Germany, the Stalinist history of science only achieved circulation because it was sanctioned by the political system. When the system changed, history changed too.

'External' legitimations need not serve direct political functions but may, for example, refer to religious views. A history of science is not ideological because it is written from a religious point of view, but it becomes so if the main purpose is the legitimation of a particular religion. Obviously, the desire to prove the case of atheism may result in histories that are no less ideological.[6]

A distinctive example of what might be called 'Christian historiography of science' can be found in Stanley Jaki, a prominent, widely acclaimed historian of science. In a series of works Jaki has carried the views of Duhem further and has firmly maintained that science is exclusively the result of the Christian faith of the Middle Ages. But whereas Duhem stressed that the validity of his view of the development of science was independent of his Catholic faith and not a particular Catholic–Christian view, Jaki goes much further. According to Jaki, only sincere Christians who have realized that the Bible is the Word of God, can really understand the history of science.[7] Historians who have a different view of the matters to Jaki and Duhem are dismissed as being blinded by anti-clerical bias and other forms of lack of orthodoxy. Thus, Mach was unable to understand the rise of science 'precisely because of his hatred of Gospel and Christianity . . . this is a principle reason why Mach could not have become a historian of science even if he had wanted to'. Sarton's lukewarm attitude towards Duhem is explained by 'the freemason Sarton's deep-rooted anticlericalism coupled with his dogmatic socialism born out of a youthful cavorting with Marxism'. As for Whewell, who at least was not anti-clerical (Whewell was a clergyman), the weakness in his *History of the Inductive Sciences* lay in the fact that 'it provided no role for the Word speaking from the Gospel in intellectual history'.[8] Etc.

Nationalistic or patriotic history of science should be distinguished from the study of national sciences. National cultures and national political and economical systems are major determinants of the style and development of modern science and hence it would be unnatural not to study these aspects. National science studies have flourished in recent time, in particular among American historians.[9] In this kind of history of science nationalistic tendencies ought naturally to be included, but only to the extent that they have actually played a part in the historical development, not via an interpretation of the historical materials.

History of science has its own 'imperialism' that partly reflects the fact that viewed historically and socially science is almost purely a western phenomenon, concentrated on a few, rich countries. While science may be international, history of science is not. The preoccupation of the history of science profession with the Great Powers of the western hemisphere does not only reflect the importance of these countries in the development of science. To some extent, at least, it also reflects the present economic and scientific strength of these countries. It is only in recent years that any interest has begun to be taken in the scientific developments that stem from, or have been carried over to, the non-European cultures.[10] The feeling that the dominant, so-called international history of science has passed over countries that are small, isolated or for some other reason find themselves on the periphery of the Learned Republic is widespread.[11] It is not entirely unfounded.

History of science can function ideologically in a different way than by being used externally for political, religious or national glorification. Namely, by providing a mytho–historical basis for scientists' conception of their discipline and their own role in its development. This kind of history is internal, directed to the scientists or the novices in the field and usually produced by the scientists themselves. Scientists are not merely the passive objects of the history of science; they are also active consumers and producers of history of science.

In the terminology of Kuhn, one can say that a form of disciplinary history appears as a necessary part of the paradigm of a scientific discipline. The historical element appears especially in the exemplars, the shared standard types of concrete solutions to problems that serve as models of how the speciality ought to be carried out. Exemplars are mainly taken from the history of science. Knowledge of the historical exemplars and of the father figures of the discipline or institution is an important part of the process of socialization that the scientist has to go through in order to be counted as one of the practitioners of the discipline. The history of science that forms part of the discipline's or institution's tradition constitutes the scientist's self-understanding and cultural tradition: how his subject has developed, which areas and methods are of value, who the founders and authorities of the discipline are, what its higher aims are, and so on. This kind of institutionalized history of science has been called the scientists' 'working history'.[12] This is not merely a retrospective history, but a practical, forward-looking history that gives instructions on the practice to be followed by those who work in the discipline or want to join it.

Because of its practical function in the sociology of the scientific community, the working history is mythical. The extent to which it gives a true account of the development is irrelevant. The working history constitutes a quasi-historical reference frame with implications relating to disciplinary policies that are common to the scientific community. It is of the same type as the national or religious history that gives a people a common national background or a common identity to a religious community.

The working history is essentially static and serves a socializing function. It is the kind of history that marks periods of normal science, in which there is no disagreement about the discipline's foundation. In cases of paradigmatic shifts the working history becomes insufficient and is often challenged by new disciplinary histories which are intended to re-define the boundaries and methods of the discipline. New histories may be constructed either to stipulate a revolution not yet accomplished, or to revise the practitioners' conception of the discipline after a revolution has already occurred.[13] Disciplinary histories that legitimate intended revolutionary changes will normally be combated by more conservative versions of the history.

Occasionally, scientists use history by referring in their works to recognized authorities and exemplars. If a work, by using such references, can be shown to belong to a historical research tradition to which a great deal of prestige is attached, part of that prestige will be communicated to the work in question; or new, unorthodox ideas can be made to appear more revolutionary by being placed as contrary to an orthodox historical tradition. On the other hand, new ideas have often been criticized by means of quasi-historical arguments; either by attacking them for being heretical in relation to the accepted orthodoxy or by asserting that they are not at all new, but merely repetitions of ideas that have appeared earlier in the history of science.

The clearest case of a historical model that has been able to govern research is probably the massive influence of the Newtonian paradigm in the 19th century. Newton – not the authentic but a half-mythical Newton – was an authority during that period who was often referred to for endorsement of new theories or for a criticism of them. Not only in physics, but also in chemistry, biology and the earth sciences was Newton's authority frequently invoked. When Thomas Young (1773–1829) proposed a new theory of light in which light was considered to be an undulation in an ubiquitous ether, he went to a great deal of trouble to present it as a natural extension of Newton's own ideas. According to these ideas, as usually understood, light is a stream of subtle particles, not a wave or vibration phenomenon. However, Newton had also speculated on light as ether vibrations and Young was in fact indebted to these speculations. 'A more extensive examination of Newton's various writings', concluded Young, 'has shown me that he was in reality the first that suggested such a theory as I shall endeavour to maintain; that his own opinions varied less from this theory than is now almost universally supposed.'[14] Thus Young attempted to revise the history of optics in order to help him stipulate the changes he believed should be made in the discipline. However, Young's theory made little impact on the scientific community at the time. Earlier vibration theorists, such as Benjamin Franklin, were often criticized on quasi-historical grounds, for challenging what was believed to be Newton's optics. Young too was criticized for restating old Cartesian hypotheses which conflicted with what the great Newton had taught.

At the end of the 1850s, the law of energy conservation was generally accepted and recognized to be one of the cornerstones of science. However, since the concept of energy is not to be found at all in Newton, his glory could hardly be made to shine on the energy law. Neither was it, at that time, necessary to legitimate the victorious principle of energy conservation. Even so, some scientists felt that the prestige of the Newtonian tradition could only remain unbroken if continuity in history could be restored also in this case. Accordingly, Tait and Thomson, two of Victorian England's most prominent scientists, reinterpreted passages in Newton's *Principia* in such a way that Newton appeared as the true originator of the principle of energy conservation. In this way the discovery of energy conservation could be regarded as a fulfilment of Newton's inspired anticipation.[15]

Scientists often use versions of disciplinary history in order to justify the originality of their own contributions. In some cases this has happened implicitly, by conspicuously omitting history altogether. One such example is Lavoisier's epoch-making *Traité Elémentaire de Chimie*, published in 1789. Lavoisier was acutely aware of his mission as a revolutionary and wanted to present his work as an entirely new foundation for chemistry. In order to stress that scientific chemistry only came into existence with him, Lavoisier ignored completely the works of earlier chemists. To mention their works, even if to criticize them, would, Lavoisier reasoned, diminish his claim of absolute originality. While Lavoisier's work did not use history, it did produce a radically revised historiography of chemistry.[16]

In Lyell's *Principles of Geology* (1830), another of the classic works of science, history was not ignored. On the contrary, Lyell's work was prefaced with four chapters in which the history of the earth sciences was discussed at length. This version of the history of geology achieved an authority, by virtue of Lyell's success, that survived for generations. Historians and geologists accepted Lyell's history as definitive and composed their picture of the development of geology in accordance with what Lyell had written. Modern, more critical historians have demonstrated the mythical character of this tradition and regard Lyell's history as predominantly a piece of self-promotion.[17] The main message in Lyell's historical preface was that until 1830 geology was at a primitive, unscientific stage of development; it was *Principles of Geology* that broke away from earlier prejudice and started the era of what would become scientific geology. As Lavoisier had succeeded in becoming the Newton of chemistry, Lyell wanted to establish himself as the

Newton of geology. In order to put his messa ṛe firmly across, Lyell produced a distorted history that consisted o a few great scientists with fundamentally inadequate or wrong ideas about the evolution of the earth. Lyell's tactics consisted partly in inventing contradictions in geology that did not really exist, and partly in making views that were rivals to his own seem ridiculous. His masterly propaganda on his own behalf bore fruit for more than a century.

In a much-quoted piece of advice, Einstein once said: 'If you want to find out anything from the theoretical physicists about the methods they use, I advise you to stick closely to one principle: don't listen to their words, fix your attention to their deeds.'[18] This is generally sound advice which need not be restricted to theoretical physics. But it should not be taken to imply that what counts is merely published scientific contributions. From a historical point of view, scientists' words, their reflective and retrospective accounts of what is going on, cannot be sharply separated from their deeds. The historical narratives that scientists produce do not reflect their scientific contributions but rather their images of themselves and their science. To the historian, scientists' more or less amateurish accounts of the history of science are valuable source materials as regards the scientists' personal attitudes and images.

Let us briefly look at Einstein's attitude to the history of science. Like so many other scientists, Einstein made extensive use of the history of science and he developed his own view of how the history of physics ought to be presented.[19] According to this view it is the task of history of science to reconstruct the exemplary concepts and principles that can serve the purpose of structuring the development of science in a meaningful way. Einstein's own semi-historical works illustrate this programme. They are exemplary, not factual, history. They concentrate on conceptual themes (such as the field concept) that are structured and selected in an idealized way in order to reveal connections that do not appear in the factual history. Accordingly, Einstein often organized the historical materials without feeling himself bound by its chronological order. Einstein did not, apparently, think very highly of historians of science, about whom he declared that they 'are philologists and do not comprehend what physicists are aiming at, how they thought and wrestled with their problems'.[20] He found his ideal of a history of science in the works of Ernst Mach, a physicist rather than a historian. It should be obvious that there is no reason to transfer

Einstein's authority from physics to history. That his views are, nevertheless, of interest is because they are those of one of the giants of science.

In 1912, three German physicists, Max Laue, Walter Friedrich and Paul Knipping, discovered in Munich that X-rays produce a diffraction pattern if transmitted through a crystal. This important discovery proved not only the wave nature of X-rays but also the lattice structure of crystals. Like other important discoveries, the discovery of X-ray diffraction has been the subject of much quasi-historical interest, resulting in an official, disciplinary working history. This history has been criticized by the historian Paul Forman who regards it as a myth.[21] The exemplary importance of Forman's analysis warrants a more comprehensive treatment in the present context.

The background is that since 1912 X-ray crystallography has developed into a separate discipline with its own social structure in the form of periodicals, congresses, international union and professional networks; included in this disciplinary structure is also a shared mythology. The community of X-ray crystallographers has institutionalized its history, in particular in the form of festschrifts and recollections of leading members of the community. The history of the event that created the disciplinary tradition is based on the retrospective thoughts of Laue, Ewald and Bragg, all of whom were among the founders of X-ray crystallography. The official creation myth has been concerned with the question 'why was X-ray diffraction discovered in 1912 in Munich?' Its answer is that the discovery was conditioned by two factors:

(1) Acceptance of, and interest in, the lattice structure of crystals.

(2) Acceptance of, and interest in, the wave-like nature of X-rays.

The discovery would naturally be made in the place where these conditions were fulfilled. The reason why that place happened to be Munich was that, according to Ewald and Laue,

(1') The idea of crystal lattices was an outsider theory in 1912, rejected everywhere but Munich; in other places the physicists were not interested in or familiar with crystallography.

(2') The idea that X-rays were waves had a strong following in Munich, while in most other places X-rays were regarded as pulses or currents of particles.

It is noteworthy that in this case the official historiography does not, as so often happens, explain the pioneering discovery as resulting from the genius of the discoverer, but as a result of the professional environment.

Forman now claims that the official creation historiography is a myth, since he shows that (1') and (2') have no basis in fact. In 1912 both crystallography and the wave-like nature of X-rays were accepted ideas that interested many physicists in Europe. So (1) and (2) cannot be satisfactory explanations of why it should have been Laue, Knipping and Friedrich who discovered X-ray diffraction. The reason why Forman calls (1') and (2') myths rather than mere mistakes is that he believes that they fulfil a particular legitimating function for the scientists involved in X-ray crystallography, viz. 'to strengthen tradition and endow it with a greater prestige by tracing it back to a higher, better, more supernatural reality of initial events'.[22] Forman adopts what he calls an anthropological perspective. He believes that, essentially, a modern scientific community can be analysed using the same sociological and psychological methods as those used by anthropologists when studying primitive tribes.[23]

The techniques involved in the formation of myths, according to Forman, are connected with, firstly, erecting barriers that the hero of the myth has to overcome, and, secondly, presenting the discovery as morally exemplary, i.e. methodologically correct. In the case of X-ray diffraction, however, the mythic hero is not an individual but an environment. Forman says:[24]

> The myth will the better serve its social function the more numerous and difficult are the obstacles to be surmounted by the mythic hero. It is in this way that we may understand the increasingly categorical assertions of the disreputability of the space lattice theory. So also may we understand the assertion that the first experiments involved exposures of many hours, when it is almost certain that in fact they did not last thirty minutes. The physicist, however, demands something more from his myths than does the savage – they are to be consonant with what he knows to be good physics, and they are to be internally consistent, even if implausible. . . . An opinion which historically was beyond the fringe, which was decidedly unorthodox and which, for one reason or another, the orthodox scientists regarded as dangerous, becomes in the myth the dominant, orthodox opinion in the science. The myth then has that threatening 'widespread' opinion being overthrown by the mythicized event or discovery.

Forman's critique of the historiography stemming from 'the culture of the fraternity of crystallographers' has been answered by P.P.Ewald, one of the patriarchs of this fraternity.[25] Ewald, in my opinion rightly, warns the historian against creating myths in order to have some myths to puncture. It is not only scientists who are capable of mythicizing history. The following edited dialogue illustrates the basic differences of opinion between the critical historian and the scientist.[26]

F: 'Myths and anecdotes – a species of minor myths – have important, and perhaps even legitimate, functions in contemporary science . . . but because they purport to be historical, myths and anecdotes are subversive to history.'

E: 'Why does the historian require myths for the preservation of the identity of the crystallographic "clan"? Would this group not also find its identity, as it was the case, in common interest, common methods and cognate problems of research, and common experience of the development of their field – that is, on factual, not mythological grounds?'

F: '. . . the scientist, *qua* scientist, places no value upon historical fact; history is wholly subordinate to the needs of the present, and indeed only survives to such extent, and in such forms, as serves present needs. . . . so long as he avoids questions of "priority", his colleagues are not obliged – indeed, not even entitled – to criticize his exposition on the grounds that the historical facts are stated incorrectly.'

E: 'True, scientists are not trained historians; instead, they have had the personal experience of growing with their subject and knowing of the motives prevalent during the period of growth. Is the historian really entitled to disqualify his descriptions of what happened and what the motives were as being myths and anecdotes? And this only because myths play an important role in primitive society! Can he properly recognize motives out of the pages of journals? Or evaluate facts without being influenced by his *a posteriori* knowledge of what *should* have been known, or done, or thought?'

Ewald raises some questions here that are central indeed to the historiography of science. In my opinion, Forman is justified in rejecting the historical acccounts of actively involved scientists as witnesses to the truth; as historians are generally justified in doing in many cases (but not always or unconditionally, of course). It ought to be noted that Forman does not actually accuse the scientists of only producing myths. It is personified science, the scientist *qua* scientist, to whom historical reality is irrelevant. In practice, living scientists are never only scientists and may well be excellent historians.

Admittedly, the historian cannot 'recognize motives out of the pages of journals'; but he can point out inconsistencies, examine unpublished materials and in other ways use the methods of historical criticism to uncover motives. As Ewald points out, the historian cannot completely avoid being influenced by his *post factum* knowledge. But this is even truer of those scientists who comment on earlier research with which they have been involved. By using a diachronic perspective the historian can at least minimize the distortion that tends to lie in his placement in time. Furthermore, the historian will rarely, unlike the scientist, be personally involved in the history in question and is therefore better able to produce an impartial analysis.

At any rate it appears that in almost all those cases where scientists have given historical accounts of research done by themselves or their colleagues, the historian can point to mistakes or inadequacies (see chapter 13). 'The personal experience of growing with their subject and knowing of the motives prevalent during the growth' does not make the scientists witnesses to the truth. On the other hand it does not automatically disqualify their statements as myths either.

11

Sources

A source is an objectively given, material item from the past, created by human beings; a letter, for example, or a clay pot. But this item is not in itself a source. It can be called a relic of the past or a source object. If the relic is to achieve the status of source-material it must be evidence from the past, it must tell us something about it. The relic must be capable of being utilized to give some of the information that it contains in a latent form. It is the historian who turns the relic into a source through his interpretation. By posing questions to it from particular hypotheses (that do not themselves need to have any documentary basis) the historian forces the source to disclose information. Unlike the relic, the source is not, as a source, a material item, but has to be regarded as information that has been released. The information disclosed by the source, and in that sense the source itself, becomes an interplay between the source object and the historian, a meeting between past and present. It follows from this that while the source object is fixed, the very same source can disclose different and possibly conflicting information.

In previous chapters we have seen that source materials are not given once and for all but that they originate in the dialectical process between the relics of the past and the interpretations of the present. History of science sources are no exception to this. The philologist and historian Julius Ruska (1867–1949) described the relationship as follows:[1]

> The history of the sciences will continue to be dependent on the sources that are at its disposal at the time, but the correct evaluation and use of the sources will, in turn, depend on the historian's ability to carry out historical criticism. Like science itself, the presentation of its history is a process that never ends.

Some sources are accounts of the past that have been written with the object of telling something about the present that once was; either directed at contemporaries or – more rarely – at future generations. Such sources as these that intentionally provide evidence, are often called eloquent or symbolic sources. In contrast to these are the 'dumb' or non-symbolic sources that only give information unintentionally or unwillingly. Both symbolic and non-symbolic sources are created by human beings and the dividing line between them is not very sharp. Letters and other written documents are typical symbolic sources. Unlike non-symbolic sources they can contain information that is of a normative kind, for example evaluations of the situation existing at the time when they are written. It is mainly the symbolic sources that offer problems in connection with the critical analysis of sources. The sources that are of most relevance to the history of science belong, in the main, to this group. A retort from Liebig's laboratory is a non-symbolic source; notebooks containing records from the laboratory are symbolic sources.

Among other things, the objective of source analysis is to determine the independence and reliability of sources. In this connection it is usual to distinguish between *primary* and *secondary* sources. By primary source we mean a source that stems from the time about which it discloses information and as such has a direct connection with the historical reality (in a chronological sense, not necessarily as far as reliability is concerned). A secondary source stems from a later period than the one for which it is a source, and builds on earlier, primary sources. The distinction between primary and secondary sources is only meaningful when applied to symbolic sources. Moreover, the distinction is not a sharp one. Since a source is only a source in a specific historical context, the same source object can be both a primary or secondary source according to what it is used for. Duhem's *la théorie physique* will be a useful secondary source for the historian who wishes to study the history of the theories of gravitation; it will be a fine primary source for the historian who wishes to investigate positivist views of science at the turn of the century.

What then are the typical primary sources that are to be found in history of science? It is not possible to make an exhaustive list but the most important sources are the following:

1a Letters
1b Diaries, laboratory journals
1c Notebooks, private notes
1d Manuscripts and rough drafts of scientific works
2a Protocols and minute books of scientific institutions
2b Reports and accounts from scientific institutions
2c Applications for posts, advertisements of posts and evalua-
    tions of the applicants; documents concerning admission to
    learned societies and similar institutions
2d Applications for patents and official patent statements
3a Unpublished theses; award-winning works, dissertations, etc.
3b Preprints
3c Published scientific articles and books (or papyri, inscriptions,
    etc)
4a Reviews
4b Textbooks, exam papers, lecture notes
4c Handbooks, tables, manuals
5a Autobiographies, memoirs
5b Films, illustrations, maps, photographs, television programmes
5c Tapes, radio programmes
5d Interviews, questionnaires
6a Official reports, ministerial memoranda, legal documents
6b Plans and sales lists from instrument-makers, science pub-
    lishers and other science related firms
7a Non-scientific books and articles
7b Newspapers
8a Libraries
8b Bibliographies

In the list an attempt has been made to divide the sources into
groups in accordance with the following idea: the sources that are
placed in 1, 3, and partly in 4 are connected with scientific work
seen as a creative, intellectual activity. Groups 2 and 6 have to do
with the social and institutional environment of science. The
sources placed in group 5 touch on varied aspects of science, mainly
of a non-technical nature. Group 7 illustrates just how diverse
printed sources can be. It is especially in connection with the social
and cultural aspects of science that the possible information will
be spread out over many different sources that otherwise do not
relate to science: novels, poems, magazines, newspapers, etc.

Another way of classifying sources has been suggested by Ottar Dahl, who makes a distinction between personal and institutional sources that can both be either public or 'confidential' (non-public).[2] In a modified form, using the descriptions given above, the scheme can take the following form:

	personal sources	*institutional sources*
confidential	1a,1b,1c,1d,5d (2c,2d)	2a,2b
'semi-public'	3a,3b,8a	6b
public	3c,4c,5a,7a,7b (4a,4b)	2b,6a,8b

The sources that have so far been mentioned have been of a symbolic kind, containing written information (with the exception of 5b and 5c). The source objects are made of paper or similar materials. But non-symbolic primary sources also exist that are of importance to history of science:

9a Buildings, laboratories
9b Instruments, machines, apparatus
9c Concrete models, plates and tablets
9d Chemicals, herbaria, natural history collections

Compared with sources on paper, such sources as those above are few and their existence is fortuitous; but when they do exist they can give valuable information about the experimental and technical aspects of science that can easily be underestimated if the historian only relies on written sources. Sources of type 9 are of special interest to the historian of technology. While the written sources are normally kept in archives, museums are the natural place to find sources of type 9.

Secondary sources are less diverse than primary sources. They often consist of the following categories:

10 Memorial volumes, obituaries
11 Biographies (not contemporary)
12 Retrospective reflections
13 History of science works
14 Other historical works

There will be no attempt here to go through all the above-mentioned types of source systematically. David Knight has given a thorough account of history of science sources, and the reader is referred to this.[3] In what follows I will confine myself to commenting on some of the sources.

Sources 1a–1d are the most direct expressions of the actual scientific process and for that reason are of special interest. Since the sources are not intended for a public they can usually be regarded as evidence with a large degree of reliability. They will not only be reliable evidence of methods and ways of thinking, but also of experimental data that normally only appears in the finished publication in a condensed, edited and possibly manipulated form. For this reason, laboratory journals and similar objects are invaluable sources when reconstructing the course of events in history of science.[4] In recent years, much has been done to preserve and file letters, manuscripts, notebooks and other things that concern modern research.[5]

The central importance of the non-public primary sources is connected with the important distinction between the so-called 'context of discovery' and the 'context of justification'.[6] While the former refers to the procedures used to produce scientific knowledge, the latter refers to criteria of acceptability for such knowledge. The intellectual historian, by definition as it were, will be committed to the context of discovery. As far as the uncovering of contexts of concrete discoveries is concerned, published primary sources are not the most reliable evidence. It is only in rare cases that publications give information about the authentic process of research.

One such case is Kepler's *Astronomia Nova* (1609), the preface of which states: 'What matters to me, is not merely to impart to the reader what I have to say, but above all to convey to him the reasons, subterfuges, and lucky hazards which led me to my discoveries.'[7] But Kepler is an exception, both from the norm that existed in Kepler's own time and in particular from what it has been in later periods. It is worth noting, however, that the existing publication norm with its sharp distinction between the contexts of discovery and justification is not a necessary part of scientific discourse. In some phases of scientific development, denoted 'concrete' by Caneva, it was part of scientific integrity to present evidence and thought processes just as they actually came to the scientist. This standard was not merely considered to be good tone but was also regarded as a criterion of truth. The following quotation from 1821 is typical of this 'concrete' standard of science:[8]

> Since the view sketched . . . *really* arose with me in the order of the
> investigations as expressed in the three sections, I considered myself,

as it were, obliged to maintain the same order of presentation, for it is far easier to uncover the paralogisms of a theory when one knows precisely the thought process that led or misled its author.

The 'concrete' ideal of sciences disappeared in the middle of the last century, when it was finally superseded by the 'abstract' standard that has since dominated scientific publications.

The value of non-public sources depends on the perspective and interests of the historian. As far as the cognitive aspects of science are concerned, non-public sources will have an absolute priority. But this does not apply if the historical interest is focused on science as a social phenomenon, for example. With this perspective, notebooks, laboratory journals and manuscripts will be largely irrelevant. The very fact that these sources are private means that they have little to say about the social history of science. A manuscript only known to the author himself cannot have had any influence on the social development of science (although it can reflect the development). The social historian will therefore be justified in focusing on different sources to those of the cognitively orientated historian, especially on the public and institutional sources.

In general, social history requires a more complex and diverse approach than does intellectual history. For example, in dealing with the development of a particular scientific field the social historian will have to examine not only the actors, the scientists, but also their audience in a wide sense. For this purpose the sources in group 7 will often be relevant. Manufacturers of scientific apparatus, firms supplying chemicals and science publishing firms constitute an important, though often neglected, factor in the development of science. The sources related to the commercial aspects of science are different from other sources and may easily be overlooked (group 6b).

While the sources in group 3 constitute what might be called the research fronts of science, the sources in group 4 will not reveal much about creative research. But textbooks will be central source materials for achieving an understanding of the established stages of development and paradigmatic basis of a scientific discipline. The same is true, perhaps even more so, of handbooks and monographs. Textbooks are condensed expositions of the authorized corpus of knowledge of a discipline and give us information about the status of the subject at any one particular time. Textbooks and similar sources thus give the norm from which it will be reasonable to evaluate scientific contributions. By discovering this norm one is able to avoid the likely mistake of identifying frontier knowledge as generally accepted knowledge, of confusing the top of the iceberg with the iceberg itself. New knowledge is not disseminated instantaneously; just because a discovery is made in one particular year it will not spread immediately and achieve acceptance no matter how true and important we, today, can see that the discovery was.

Although textbooks do not themselves form part of the active research front, they will also be of interest as sources of pioneering science; namely as the literature that young scientists-to-be have been acquainted with and which has, perhaps, for that reason played a part in their later discoveries. Thus, in an attempt to trace the sources that inspired Einstein to make his theory of relativity, Holton has argued that the most important single source was an otherwise forgotten textbook on electrodynamics by the German physicist August Föppl.[9]

Handbooks and scientific encyclopedias have a similar function as sources to that of textbooks:they will often provide an authoritative expression of what was known at a particular time. Reviews and abstracts that can be found in periodicals and bibliographical yearbooks are good sources for evaluating how a scientific work was received by a particular person or in a particular milieu. In reviews the style is often freer and the reviewer expresses his opinion in a more direct way than in scientific publications. Reviews are particularly important sources in connection with methodological controversies, priority conflicts and similar problems. For example, when Alfred Wegener proposed his theory of continental drift it was rejected by the majority of geologists. The best way to judge the intensity of opposition that Wegener faced is through the study of symposia and review articles. According to one reviewer, 'Wegener . . . is not seeking truth; he is advocating a cause, and is blind to every fact and argument that tells against it.'[10]

The kind of scientists who are made the objects of sources of type 5 are usually famous scientists whose discoveries took place many years before the time the source originates from. Accounts written by the scientists who were actually involved with the discovery belong to group 12 but also have something of the nature of primary sources. They will often have great authentic value and in many cases they will be the only sources of knowledge. There are personal questions that only the scientist can answer and which are difficult to test on the basis of other knowledge. However, reminiscences and autobiographies are not always reliable and should be used critically (see chapter 13).

Visual sources will only rarely be of any great interest with reference to the origins of scientific discoveries. The working scientist is not followed around by a film or TV crew during his creative moments. Nevertheless, visual sources are very important, in particular in relation to the information they can impart about the general conception of science in the past. Medieval illustrations of the human body may, if properly interpreted, yield information about the medical knowledge of the Middle Ages that cannot be found in any text. Illustrations cover a wide spectrum: maps, diagrams of apparatus, portraits, plates of natural history objects, illustrations of models and analogies, graphical representations, etc. Whatever their sort, visual sources are always designed to accompany a text and should be examined as such; and yet illustrations may surpass the text and acquire a life of their own, transcending the barrier between science and art. The anatomical illustrations of Leonardo de Vinci and the zoological illustrations of Albrecht Dürer are well-known examples. Analyses of old pictures with scientific motifs require the historian of science to work as a historian of art. In some cases pictures are an important source of knowledge about the material base of early science, depictions of laboratories, dissecting scenes and apparatus, for example.[11] Diagrams and technical drawings are indispensable sources in most areas of history of technology.

The study of what scientists have read can give important information about their background in general and in particular about other scientists who have influenced them. If it can be documented that a scientist had read a particular work before he made his discovery, this work might possibly be of some significance for the discovery, even if the scientist himself does not refer to it.

In the very few cases where the private library of a scientist still exists in reasonably good condition, or where the library can be reconstructed, the historian will have a unique opportunity to form a picture of the life of the scientist.[12] But one obviously cannot form any conclusions merely on the basis of the fact that a scientist X owned a copy of a work written by Y. There can be many reasons why X had a copy of Y's book. He might have received a copy from Y, for example, without having read it. If one wants to establish whether X was influenced by Y's book in his work, one has to ask whether X actually read the book. Have the pages been cut? Are there any 'dog-ears' or other signs of use? Has it been annotated? Has X made any references to the book? When did X read it? And so on. It is from knowledge of the books owned by Newton, among other things, that his interest in alchemy can be documented (cf. Chapter 2). When someone leaves over 100 volumes that can be classified as alchemy when he dies, one must at least conclude that the person in question was seriously concerned with alchemy.

Obituaries and similar memorial articles are valuable but problematic sources. They are problematic because the purpose of the obituary is not primarily to give reliable historical information but to glorify the character and life of the person who has died. Obituaries are almost always uncritical or at least favourable portrayals of a life. Furthermore, they are usually written by the deceased's colleagues or students, for whom the obituary will tend to serve as a link in the working history of their disciplinary tradition. In short, obituaries are examples of the kind of mythical historiography that we discussed in Chapter 10.

Going through bibliographies will often be a good introduction to a history of science work. Bibliographies can be arranged in many different ways, a bibliography of a subject or discipline in a certain period, for example, or bibliographies of individual scientists. For many scientists, there are more or less complete bibliographies in existence, in some cases covering more than 1,000 published works. Complete bibliographies ought not only to contain original scientific works but also works of a non-technical nature, plus review articles, information about translations into foreign languages, numbers published and number of editions. Bibliographies of secondary literature are of equally great practical value. The current bibliographies of *Isis* that encompass all recent history of science publications are a necessary tool.[13]

The process of historical research begins with a problem situation. The historian chooses this problem situation in connection with his desire to deal with a particular subject. He formulates questions on the subject, builds up an idea of what it is that he wishes to know. These questions will then naturally lead to particular sources that might possibly be able to answer the questions and that will probably lead to new ones. The original problem situation will be transformed through the research process, partly as a consequence of the study of sources.

The first stage will be to track down and identify the sources of relevance to the problem that has been defined. This can be a difficult task, depending on the nature of the subject and its position in time. It will often be a good idea to start with the secondary sources, in particular with the works that other historians have written about the same or similar subjects. In this way one can save on much time-consuming reference work and relatively quickly achieve an overview of the source materials that will need closer investigation. No matter how thoroughly he searches for source materials, however, the historian will never succeed in basing a study on *all* relevant sources. It is impossible to know whether relevant information can be found in sources that have not been consulted; and irrelevant sources might, nevertheless, turn out to be relevant. It is the interpretation of the source that determines its relevance for the question posed by the historian.

Once the historian has chosen his sources, he ought, in principle, to examine their authenticity. In other words, he ought to be alert to the possibility that they might have been faked. I only know of one single case of actual forging in the history of science, but this is so extensive and grotesque that it deserves mentioning.[14] It concerns a certain Vrain-Lucas, who manufactured thousands of false historical documents in the 1860s, including letters from Luther, Galileo and Newton (not to mention Pontius Pilate and Mary Magdalene – all of them in French!). Among the documents of the inventive Vrain–Lucas was an exchange of letters between Newton (aged 11) and Pascal which revealed that the latter was the true discoverer of the law of gravitation. The most disturbing fact about the Vrain–Lucas affair, perhaps, is the fact that his home-made sources were taken seriously by several French scientists, whose patriotism and scientific vanity were given a boost by the false documents.

Historical criticism is the process by which sources are analysed critically with a view to establishing their authenticity and reliability.[15] The aim is to evaluate how close to historical reality the sources lie, since one makes a prior assumption that no source gives an exact reflection of the past but can only be more or less complete signals from the past. It is important to establish whether the source is authentic, whether its dating and information about location is correct, whether its assumed author is its real author, and so on. Sources that are not authentic in this sense need not be forgeries (– but forged sources are never authentic). There can be many reasons why the immediate information disclosed by the source does not reveal its actual origins. Sarton illustrated how even an apparently reliable primary source can give false information. On the colophon of an early printed edition of Ptolemy's work on geography (*cosmografia*) the edition is dated MCCCCLXII, that is 1462. But it is almost certain that this date is wrong and that the book was not printed until 15 years later.[16]

The central investigation in the analysis of sources is the one to establish the reliability of the source. Does the source represent historical reality? How reliable is its information? As has been mentioned, there can be many reasons for not immediately accepting the information in a primary source without further enquiry. The information given is typically the author's version of reality and must always be evaluated in the context in which it occurs. One has to analyse the motives that the author had for writing what he did, establish the reason behind the source. To whom was the source originally addressed? Under what circumstances was it written? And first and foremost one has to compare the information in the source with other evidence concerning the event the source deals with; partly compare with other sources and partly scrutinize the contents using what is generally known about the subject and the time. The misdating of the above mentioned edition of Ptolemy's Geography was not the result of a conscious action, for example; the publisher or printer could hardly have had any motive for dating the book 1462, which must have been the result of a simple misprint. The mistake can be recognized as such by comparisons with other editions of Ptolemy's book (the first being from 1475) and from contemporary commentaries on the work.

As we shall show in the next chapter, the heart of all criticism of sources is the comparison of different evidence. Let us think of a particular occurrence 0 that has possibly taken place in the past. In order to determine whether 0 is true or false, we have a series of sources that give different evidence, $E_1, E_2, E_3, \ldots$ Usually, some of this evidence will support 0 and other evidence oppose it; we can denote such evidence E^+ and E^-, respectively. There can now be different situations:

1 If 0 is clearly in conflict with established scientific knowledge one will immediately conclude that 0 did not take place and that E^+ is either false or has been misinterpreted. This also applies even when there are no sources with E^-. Note, however, that the knowledge one uses here to evaluate 0 is the established knowledge of one's time and that the evaluation can therefore never be more certain than this knowledge.

2 If only E^+ or only E^- exists, and there are no special grounds to reject or accept 0, the matter is trivial. In this kind of case one will, of course, conclude that 0 respectively did or did not take place.

3 If there is conflicting evidence, between E_1^+ and E_2^-, for example, there are two possible situations:

3a If there is further evidence ($E_3, E_4, \ldots$), one will compare this with E_1 and E_2. If $E_3 = E_3^+$, $E_4 = E_4^+$, and so on, one will conclude that E_2^- must be rejected and that 0 did in fact take place.

3b If the only evidence is E_1^+ and E_2^- the historian has to assess whether E_1 or E_2 gives the most 'plausible' or 'reasonable' account. The historian might be forced to admit that it is not possible to distinguish between the reliability of E_1 and E_2 and that knowledge of 0 can thus not be established on the basis of existing sources.

4 It could be that E_1 does not actually conflict with $E_2, E_3, \ldots$, but that E_1 is not supported by other, independent evidence. In this case E_1 will have a unique status in relation to other evidence about 0. Normally, an acceptance of E_1 will require a correspondence with other evidence. The absence of such corresponding evidence will make the total sum of evidence $E_1, E_2, E_3, \ldots$ unintelligible and incoherent. One will then conclude that E_1 must be rejected.

5 The lack of supplementary evidence cannot automatically, however, be used as grounds for rejecting E_1. Marc Bloch reminds us of this:[17]

> The reagents for the testing of evidence should not be roughly handled. Nearly all the rational principles, nearly all the experiences which guide the tests, if pushed far enough, reach their limits in contrary principles or experiences. Like any self-respecting logic, historical criticism has its contradictions or, at least, its paradoxes. . . . for a piece of evidence to be recognised as authentic, method demands that it shows a certain correspondence to the allied evidences. Were we to apply this precept literally to the letter, however,

what would become of discovery? For to speak of discovery is also to speak of surprise and dissimilarity. A science which restricted itself to stating that everything invariably happens according to expectation would hardly be either profitable or amusing.

As an example, there is documentary evidence that Leonardo da Vinci seriously concerned himself with the principles of flying and sketched plans of flying machines. This evidence does not correspond with other contemporary evidence, since Leonardo was evidently the only one who discussed the possibility of flying in the Renaissance. Nevertheless, we accept Leonardo's sketches as authentic sources, namely as an extremely original contribution that is more an expression of his genius than of the general conditions of the age.

12

Evaluation of source materials

Any evaluation of primary, published materials will involve the question of whether the text can really be attributed to the author; or how authentic an expression it is of the author's own thoughts. One cannot unquestionably assume that every word in a scientific publication is that of the author. There can be many reasons for this. It is well known, for instance, that for a long time there has been a tradition in academic institutions according to which professors, directors, head doctors and similar highly placed personnel appear as the authors of papers that have really been written by, and based on, the work of younger researchers. Furthermore, one must be aware of the fact that published sources have always to a certain extent been filtered through the apparatus of publication; that editors of periodicals, for example, might have changed the paper written by the author, sometimes quite a lot, and not necessarily with the consent of the author. In earlier times it was often the right or even duty of the editor to modify the material that was submitted, rather freely. In cases like this one cannot use the published source as a reliable expression of the exact views of the author. Today, scientific articles are criticized and edited by referees; the version that is published is often a second or third version of the original manuscript and thus not a suitable source of detailed information about the views of the author. Rough sketches and earlier unpublished versions of manuscripts will be much more suitable for this purpose.

The problems connected with anachronical and diachronical historiography respectively, come out into the open when one analyses sources written in a language that is significantly different from one's own. Superficially, the ideal to aim at might appear to be an exact translation that rigorously reproduces the form, contents and sense of the original. On closer inspection, such an ideal will be seen to be worthless, in fact it will not result in a translation at all. The only way in which a historical text can be reproduced with absolute precision is to reproduce the original as it is *in extenso*. This is like saying that the most exact map that it is possible to make is a totally true-to-nature 1:1 reproduction of the landscape that cannot be distinguished from the landscape itself. Such a map would obviously not serve any purpose. The reason for making a translation is to transform the information in the source from the past to the present, to make it understandable in a contemporary context. Unlike a photographic reproduction, an actual translation will involve interpretation and evaluation. 'Every good translation is an *interpretation* of the original text', as Popper has pointed out. 'I would even go so far as to say that every good translation of a nontrivial text must be a theoretical reconstruction.'[1]

The question of translatability in history of science is closely bound up with fundamental questions of theory of science. These especially touch on translatability between different theories (between Aristotelian and Newtonian physics, for example), but do not essentially differ from the problems that confront the historian of science. The degree of untranslatability between different theories is one of the controversial points in modern theory of science, between Kuhn and Popper, for example.[2]

Precision is, of course, a virtue in translations. But it cannot be a goal in itself and might well militate against the clarity and information that the translation is rightly expected to bring to the reader. The reason why a certain amount of freedom and interpretation – and thereby also a certain measure of anachronisms – has to be included in translations of sources is bound up with the nature of the historical process; that it is not only a process between the source and the historian, but one which also includes the contemporary public that the historian is addressing. It is not enough for the historian to understand the sources of the past himself as a result of his studies and his empathic insight into the past. He must also be able to communicate his knowledge to a public that has not made a close study of the original sources.

But naturally, this does not mean that precision and care in translations are illusory virtues. Far from it. Within the boundaries formed by the source and the context of historical analysis, it is the plain duty of the historian to reproduce original texts as authentically as possible. This may seem to be a trivial point, hardly worth mentioning, but the fact is that even in scholarly works on history of science quotations from sources are often distorted.[3]

When reproducing extracts from sources it is self-evident that the actual quotation ought to be clearly marked off from the rest of the historian's account and that the quotation ought to be accompanied by a reference to the text from which it has been taken. It is, furthermore, a well-known fact that quotations can be misused, for in the very nature of things they are torn out of their original context. It is very easy, and often tempting, to quote a source in such a way that the quotation, in spite of it being precise, does not represent the actual contents of the source. With the use of scissors and paste, quotations from the same source can easily be used to support quite different conclusions. It is up to the honesty of the historian and his understanding of the source in its entirety to ensure that the source is not misrepresented or that there is no direct tinkering with the quotation. Let us look at an example.

In 1896 the American historian Andrew White wrote a long and learned work on the historical relationships between science and Christianity, a work that was an authority in this field for a long time. The following concerns the attitude of the Calvinists to Copernican teachings:[4]

> While Lutheranism was thus condemning the theory of the earth's movement, other branches of the Protestant Church did not remain behind. Calvin took the lead, in his *Commentary on Genesis*, by condemning all who asserted that the earth is not the centre of the universe. He clinched the matter by the usual reference to the first verse of the ninety-third Psalm, and asked, 'Who will venture to place the authority of Copernicus above that of the Holy Spirit?'

Following White, Calvin's alleged anti-Copernicanism has been a permanent part of history of science and history of ideas for generations; the quotation from Calvin used by White has been used as evidence many times, by Bertrand Russell, Will Durant, J.G. Crowther and Thomas Kuhn, among others. It turns out, however, that the quotation is a fabrication. Rosen and Hooykaas have proved that Copernicus was not mentioned in any of Calvin's works, and argue that Calvin was not an anti-Copernican at all.[5] According to Hooykaas, the standard view of Calvin as an anti-scientific, religious fanatic and fundamentalist is, on the whole, erroneous and not supported by any documentary evidence. Rosen suggests that Calvin's apparent indifference to the Copernican world picture is quite simply due to the fact that he had never heard of Copernicus. Whether or not this is the case, we can never know with any certainty, for good reasons. Calvin's failure to mention Copernicus is an example of negative documentary evidence and as such it has a different status to that of the usual positive kind of evidence. Thus, although Calvin makes no mention of Copernicus or his astronomical system, he might well, of course, have had some knowledge of these. But one has to agree with Rosen that Calvin's silence is at least puzzling if he did actually know about Copernicus's system.

Calvin's world picture was most definitely not a Copernican one but it does not reasonably follow from this that it was anti-Copernican. The alleged anti-Copernicanism requires positive documentation that Calvin did actually react to the theory of Copernicus. In view of the controversial significance that we know Copernicus came to have for theology and intellectual life in Europe, it is tempting to read Copernican ideas into any cosmological statement from the second half of the 16th century. This is apparently what White did. Calvin's commentary on the 93rd Psalm verse, does not in fact contain any polemical references to Copernicus. But White, who knew the Copernican system to be theologically controversial, read the commentary in such a way that it complied with his knowledge and expectations.

The example of Calvin and Copernicus demonstrates that false quotations can live for a long time and achieve the status of general knowledge. It is only by checking sources that such mistakes can be corrected; and even then the distorted evaluation and the false quotation can live on and give rise to erroneous historical accounts for years to come.

Meticulousness in the translation of sources is particularly important when dealing with older specialist terms and with terms whose meaning has changed in the course of time. Linguistically speaking, several of the central expressions used are the same today as they were in the past, but their meaning might have changed radically so that a direct translation in such cases would be misleading indeed. They could be technical terms such as 'force', 'fluid' and 'element', which, if translated directly and without annotation, will often result in absurdities; or they could be more general expressions such as 'experiment', 'theory' and 'science'.

In the 18th century natural philosophers considered electricity to be a kind of 'fluid'. If now the historian, eager to give a precise translation, fails to annotate the word, the innocent reader might think that the scientists of the 18th century regarded electricity as some kind of watery fluid. Which they did not. Anachronisms arise when the reader is confronted with a text. Even though an original text cannot possibly be anachronistic in itself, it will often be regarded as such when read by later generations. It is the historian's task to prevent this from happening and with this end in view he has to refer to contemporary knowledge, speak the language of the reader. Exactly what language he is to speak depends of course on his audience.

The English words 'philosophy' and 'science' have a meaning today that differs from the meaning of the words in the 17th century; this meaning, in turn, differed from the meaning of the words in the late Middle Ages. According to John Locke (1632–1704), 'natural philosophy is not capable of being made a science'.[6] How should this sentence be translated so as to render its meaning understandable to a modern reader? One possibility is to just reproduce the sentence as it stands. In which case one would avoid the danger of damaging the text, but on the other hand the meaning would then be misunderstood by most of the readers. Did Locke believe that (natural) philosophy could not be made into a science? Did he really object to making philosophy scientific? That this was not what Locke thought and that he was not talking about philosophy in our meaning of the word is revealed by his use of the term 'natural philosophy'. Around 1700 this term was used to describe the kind of natural knowledge provided by the new science, as epitomized in Newton's 'mathematical principles of natural philosophy' (*Principia*). One might therefore translate the meaning as

> 'natural philosophy [i.e. Newton's physics] is not capable of being made a science'.

But this version is mystifying, not least when one considers that Locke was particularly enthusiastic about Newton and the new empirical science. Locke certainly did not believe that Newton's physics was unscientific. In order to grasp the right meaning of the sentence the word 'science' will also have to be transcribed, since it does not refer to what we call science. In Locke the word is used in the older Aristotelian sense, in which 'science' embraces such disciplines as logic, mathematics, grammar, astronomy and similar non-experimental knowledge. It is only then that Locke's statement becomes understandable, but then it is difficult to reproduce it in the form of a quotation. One possibility is to write it as

> 'natural philosophy [i.e. Newton's physics] is not capable of being made a science'.*

where the asterisk refers to a note on the meaning of 'science'.

Problems of translation do not arise only from words that have changed their meaning in the course of time. They also arise from words that have a general meaning but that are used in quite a different, idiosyncratic way by certain scientists or groups. Moreover, they may arise from technical terms that are no longer in existence and cannot, therefore, be given a form that is both precise and understandable to a modern reader. Early chemistry and in particular alchemy are typical in this respect. The conceptual world of alchemy was so different from ours and its use of language so allegorical and mystical that it can be well-nigh impossible to give reasonable translations. These problems are an integral part of alchemy since the alchemists hoped that their works would only be understandable to the initiated. Crosland writes that [7]

> a division was continually emphasized between adepts, who were able to interpret alchemical symbolism, and the common herd of mankind, to whom alchemy was essentially mysterious. Those who had not received some guidance might find it very difficult even to recognize an allegorical description as referring to chemical processes, quite apart from being able to interpret the details of the allegory.

In the deliberately mystifying language of the alchemists, well-known words were often used in senses known only to those initiated in the alchemical fraternity. For example, the word 'water' rarely means ordinary water in alchemical texts but may describe fluid substances in general or in some cases particular fluids.

In the rest of this chapter we shall exemplify the problems of the correct use of sources more cohesively in connection with two case studies.

Dalton's atomic theory

The question of how John Dalton's (1766–1844) famous atomic theory came into being has long been discussed by historians of science, who have come up with several – at least eight – different explanations. When historical questions are problematic it is often due to the fact that there are not sufficient, and not sufficiently reliable, source materials available in which cases reconstructions have, to a large extent, to be based on speculation. This is not so with Dalton's theory. In this case most of the primary sources a historian could wish for are known, including Dalton's own statements. Unfortunately most of Dalton's apparatus and many of his letters and manuscripts were burnt in an air raid on Manchester in 1940, at a time when unexploited source materials were still in existence.

In spite of the wealth of sources and the many historical analyses, the historical explanation of Dalton's atomic theory has been a tough nut to crack, in particular because of the mutual lack of agreement in the sources available. The problem can hardly be claimed to have been finally solved today. It probably never will be.

Which are the sources that tell us about how Dalton came to think of his atomic theory? The most important ones are the following:

a. Dalton's published scientific works, especially from the period 1801–1805. No direct information about the origin of the theory is given in these.
b. Dalton's oral accounts to William Henry (1774–1836) and his son William Charles Henry (1804–1892). William Henry, a colleague and close friend of Dalton's, described a conversation he had with Dalton in 1830. In this conversation Dalton said that he had been inspired by J.B.Richter's table of chemical equivalents (1792) to put forward his theory of simple multipla in chemical compounds. W.C.Henry, a pupil of Dalton, reported in his biography *Memoirs of the Life and Scientific Researches of John Dalton* (1854) that Dalton had given the same explanation in a lecture in 1824 from which he had taken notes.
c. Thomas Thomson (1773–1852) met Dalton in August 1804 and was told that the atomic theory came into being in connection with Dalton's study of the constitution of gases (methane, ethane). Thomson reported this in 1825 and again in 1831 in his *History of Chemistry* (1830–1831). Thomson knew Dalton well, he was a respected chemist and historian and an effective champion of the Daltonian atomic theory.

   d. In 1810 Dalton gave a series of lectures for the Royal Institution in London. In Dalton's handwritten notes for one of these lectures he gives a clear and rather detailed account of the considerations that led him to the atomic theory; namely that a reading of Newton and considerations of the different sizes and weights of gaseous particles led him to the idea that atoms combine in multiple proportions. Dalton's document, dating from 1810, was discovered in 1895 by the chemists H.E.Roscoe and A.Harden. It was analysed, together with other newly discovered Dalton sources, in their book *A New View of the Origin of Dalton's Atomic Theory* (1896).

   e. Dalton kept notebooks in connection with his laboratory work. These notebooks, starting in 1802, were also found by Roscoe and Harden. They do not contain an unambiguous answer on the subject of the origin of the atomic theory, but they do tell us what Dalton thought and worked on in this critical period and help with chronological clarification.

These sources are all valuable but of different degrees of authenticity. Thus, it is obvious that b and c, based as they are on oral statements many years after the theory was created, cannot be regarded as having the same authenticity as d and e. There is no reason to believe that Thomson and the two Henrys did not report what they thought they heard Dalton say. But we cannot be certain that their (written) statements acccurately record Dalton's (oral) statements. Furthermore, in the version given by the Henrys the oral statements are from respectively 20 and 26 years after the formulation of the theory so that Dalton's statements might well have been marked by forgetfulness and rationalization after the event.

   Thomson's version, based on evidence a, has to be rejected because it is not consistent with source materials of a more primary, more authentic nature. For Dalton's notebooks reveal that his work with methane and ethane did not begin until 1804, while his table of atomic weights (which can be regarded as the direct expression of the atomic theory) can be found in notes from as early as 1803. But why did he then tell a different story to Thomson in 1804 when everything was still fresh in his memory? It could be that Dalton was talking about his newly discovered solution to the composition of carbohydrates and that Thomson has later rationalized the conversation into having been about the discovery of the atomic theory.[8] Perhaps Thomson misunderstood Dalton's reply or Dalton misunderstood Thomson's question . . . . The general unreliability of Thomson's account can also be seen in his different versions of it. In 1831 he stated that [9]

> Mr Dalton informed me that the atomic theory first occurred to him during his investigations of olefiant gas and carburetted hydrogen gases, . . .

In a book from 1825, however, Thomson wrote:[10]

> *I do not know when the ideas first occurred to him* [Dalton]. *In all probability* they struck him by degrees and were adopted in consequence of his experimental investigations. . . . *Unless my recollection fails me,* Mr Dalton's theory was originally deduced from his experiments on olefiant gas and carburetted hydrogen.

There is a big difference between this guarded account and the apparent authority in the 1831 account. Furthermore, in 1850 Thomson gave a completely different account of Dalton's theory which he then attributed to his work on the constitution of nitrous gases.

The Henry version cannot be rejected with the same assurance as the Thomson version. It was later resurrected by Guerlac, who thought that Dalton knew Richter's table of equivalences in 1803 and that this knowledge was of vital importance to the theory.[11] Guerlac was not able to give positive documentary evidence – apart from Henry's – that Dalton knew Richter's work but argued that this *might* have been the case. As has been mentioned, Dalton's own statements, as reported by the Henrys, cannot in themselves be given any convincing value, especially since they are contradicted by other sources. In 1845 Thomson said that when he met Dalton in 1804 neither of them knew about the work of Richter and that he, Thomson, was the one who later told Dalton about it. Strictly speaking, this evidence is no more reliable than that of the Henrys, although it seriously questions the Henry version. Thomson's counter-evidence alone does not entitle one to conclude, as Greenaway does, that 'we can therefore dismiss any suggestion that Dalton's speculations flowed from the systematics of Richter'.[12]

When history of science questions cannot be answered with reference to positive evidence in primary sources one has to use arguments based on reasonableness and plausibility. If one wishes to show that Dalton had no knowledge of Richter's table in 1803 one has to evaluate all the evidence in favour of this. Apart from the evidence of Thomson already touched on, there is the fact that no mention is made of any knowledge of Richter's work in Dalton's notes from 1802–1803, that is, negative evidence. Richter's name does not appear in Dalton's notebooks until 1807. Today most Dalton scholars agree that although Dalton might have known about Richter's table in 1803, it is highly unlikely that he did; and even if this were the case this possible knowledge played no part at all in the creation of his atomic theory.

In 1911, A.N. Meldrum suggested that the atomic theory had its origin in the experiments that Dalton carried out on the composition of nitrogen oxides ('nitrous gases') which he published in 1805.[13] There is no direct documentary evidence to support Meldrum's version, which is especially based on the experimental values in Dalton's notebooks and in the article from 1805. Meldrum demonstrated that the law of multiple proportions arises naturally from a contemplation of these figures, which he believed to stem from August 1803 (some of Dalton's notes are undated). Meldrum's version gives a plausible explanation of Dalton's theory and it has long been accepted; partly because it seems so natural from the point of view of modern chemistry.

Despite its plausibility, however, it cannot easily be reconciled with the sources. There are problems with the chronology and problems with Dalton's notebooks from 1802–1803. There are also problems with Dalton's experimental results, some of which may not have been based on pure experiments but modified with the help of the atomic theory.[14] In particular, Meldrum's version cannot be reconciled with Dalton's own version from 1810. How authoritative is this source? A reasonable judgement is the following, due to Leonard Nash:[15]

> . . . while no great importance needs to be attached to Dalton's oral statements, his written statements make a far stronger claim on our considerations. To be sure, Meldrum seems to have adopted the attitude that nothing said by Dalton about the origin of the chemical theory deserves the slightest credence. But from what we know of Dalton's character, it is unthinkable that he designed his statement in a wilful effort to deceive posterity. The statement appears to be a carefully considered account, and it was destined by Dalton for presentation to a distinguished audience at the Royal Institution. It was drawn up only seven years after the event; the dates cited for the creation of the first (1801) and second (1805) theories of

mixed gases are substantially correct; and there is no reason to suppose that Dalton's memory had failed. Plainly we cannot take Dalton's statement as literally exact. Yet completely to disparage it as Meldrum does . . . is to go too far.

So Dalton's account from 1810 should be highly valued as a source. It was accepted as the truth by Roscoe and Harden. According to these authors, the atomic theory was the result of Dalton's use of hypothetico–deductive thinking, not of inductive conclusions on the basis of experiments as in Thomson's and later in Meldrum's version. Even so, there is reason to reject Dalton's own account as the literal truth. It contains the statement 'this idea occurred to me in 1805' where Dalton is referring to his idea of gas mixtures as consisting of particles of different sizes from which the atomic theory is supposed to have sprung. But this is not consistent with the first table of atomic weights appearing in Dalton's notebooks from September, 1803. Dalton's specification of 1805 as the critical year could have been a simple mistake, as Roscoe and Harden suggested. But this can hardly have been the case, since the notebooks reveal that Dalton's idea about gas mixtures first appears a year after September 1803. So Roscoe and Harden's unconditional acceptance of Dalton's account from 1810 has once more to be rejected because of disagreement between this source and more reliable sources. Note that once again a plausibility argument is being used: Dalton *could* have written 1805 by mistake and he *could* have had his theory of gas mixtures in 1803 even though it does not appear in the notebooks; but it is highly implausible.

Whatever the answer may be to the question of the origin of Dalton's atomic theory it has to be based on source materials. Some of the confusion surrounding the problem may be due to the fact that one has attempted to answer a question that does not really correspond to historical reality. Namely, the question 'what event resulted in the atomic theory?' There may have been many events involved, stretching over a long period of time; the origin of the atomic theory might perhaps be better described as an evolutionary process than a sudden event. 'In one sense', writes Arnold Thackray, 'Dalton's theory had no origin, but was rather something inherited and only gradually made explicit and formalized in response to questions arising from his work on gases.'[16]

Galileo's experiments

Research on Galileo is an industry which has produced hundreds of books and thousands of articles. In the course of time, the general estimate of Galileo has undergone many changes, partly reflecting the prevailing theories of science within which it was hoped to accommodate Galileo. Almost no matter what the view of science, it is not too difficult to make Galileo appear to be an exponent of that particular view if one uses suitable quotations. We will only be looking at certain parts of Galileo's research below; not in order to find out what Galileo's methods really were, but in order to illustrate some points that have to do with the criticism of sources.[17]

For a long time people have wanted to see Galileo's success as the result of a modern empiricist, anti-authoritarian approach and especially of a new experimental method, Galileo as a refined version of Bacon. This has been the prevailing view for 300 years, in which 'Galileo's method' has been used synonymously with 'the empiricist–inductive method'. In *Dialogo* (1632) Galileo discusses in dialogue form what will happen if one allows a body to fall from the mast of a ship that is in motion. Simplicio (advocate of Aristotelianism) maintains that the body will hit the deck behind the mast while Salviati, Simplicio's debating partner and Galileo's alter ego, maintains that the body will land at the foot of the mast. During the discussion Simplicio asks if Salviati has ever carried out this experiment, which Salviati admits that he has not. 'How is this? You have not made an hundred, no nor one proof thereof, and do you so confidently affirm it for true?' asks Simplicio. When Salviati replies that he must necessarily be right, regardless of experiments, the situation now appears to be awkward for the empiricist picture of Galileo. In Stillman Drake's precise translation of the original Latin text Salviati replies:[18]

> Without experiments, I am sure that the effect will happen as I tell you, because it must happen that way . . .

When *Dialogo* was translated into English in 1661 this proud apriorist reply was too embarrassing in an England that was strongly influenced by Bacon's empiricism. To think that Galileo of all people should deny the crucial value of experiments! Salviati's reply was toned down to [19]

I am assured that the effect will ensue as I tell you; for so it is
necessary that it should . . .

The 'empiricization' of Galileo is just as marked in a modern,
much used translation of *Discorsi*, in which Galileo states that in
his investigations into movement, he has 'discovered by experiment
some properties of it which are worth knowing and which have
not hitherto been either observed or demonstrated'.[20] So here
Galileo draws direct attention to the experimental basis of his
mechanics. But, alas, it turns out that the central words 'by experiment' do not appear at all in the original text! Either deliberately
or unintentionally they have been inserted by the translator and
thus completely distort the information one can get from the quotation about Galileo's method. The distorted quotation has often
been used as an argument for Galileo's modern approach to science;
as it was by Raymond Seeger in a spirited attempt to refute the
modern apriorist view of Galileo.[21]

Galileo thus drew public attention to the fact that he (Salviati)
had never carried out the experiment with the ship's mast and that
he furthermore regarded it as superfluous. This line of reasoning
fits in with the view of Galileo held by many modern historians
of science: Galileo as the a priori working theorist greatly influenced by Platonic, Pythagorean and Archimedean ideas, who only
used experiments at most to demonstrate results that had already
been achieved. This revaluation has been especially due to the
important works of Koyré, in particular his *Études Galiléennes*
(1939), a classic in modern research on Galileo.[22] According to
Koyré Archimedes is the chief methodological model for Galileo.
Koyré argued that 'it is thought, pure unadulterated thought, and
not experience or sense-perception, as until then, that gives the
basis for the "new science" of Galileo Galilei.'[23] Unlike direct
experience, experiments played admittedly an important, positive
part during the scientific revolution but only, emphasizes Koyré,
in so far as they were subordinate to theory.

'Galileo did not really carry out this experiment,' writes
Dijksterhuis with reference to the famous story of the leaning tower
in Pisa. 'In general one always has to take stories about experiments,
by Galileo as well as his opponents, with some reserve. As a rule
they were performed only mentally or they are merely described
as possibilities.'[24] Hall repeats this judgement, in accordance with
the tradition established by Koyré: 'Many of Galileo's experiments

(or rather, appeals to experience) were rhetorical; they were not reports of events made to occur in a precise fashion.'[25] Truesdell, another expert on the history of mechanics, points out that 'Galileo was essentially a Neo-Platonic idealist rather than an empiricist'.[26]

Does one now have to accept that Galileo was a 'Neo-Platonic idealist'? Not quite. With reference to the experiment with the ship's mast it appears that Salviati's statements do not actually cover Galileo's actions. In 1624 Galileo wrote to Francesco Ingoli, secretary at the Papal printing press:[27]

> I have been twice as good a philosopher as those others because they, in saying what is the opposite of the effect, have also added the lie of their having seen this by experiment; and I have made the experiment – before which, physical reasoning had persuaded me that the effect must turn out as it indeed does.

This source must be regarded as reliable evidence of the fact that Galileo carried out the experiments that in 1632 he more than hinted that he had not carried out.[28] Although the fact remains that Galileo was convinced of the outcome of the experiment beforehand, it was no mere thought experiment. The letter to Ingoli therefore weakens the Platonic interpretation of Galileo.

A parallel case can be found in Galileo's investigation of movements of projectiles in which he demonstrated that the trajectories of projectiles are parabolic. Galileo published his argument for this in *Discorsi*, where he described it as being based on pure thought experiments. 'I mentally conceive of some movable projected on a horizontal plane, . . .' he says. So Galileo's discovery was non-experimental? If he had carried out real experiments, one would expect that he had mentioned them and not pointed out that they were hypothetical. However, Drake has unearthed some hitherto unpublished notes in which Galileo records careful experiments with balls on an inclining plane.[29] These notes, dating from 1608, prove, in the opinion of Drake, that Galileo realized from experiments that after leaving the plane the balls move parabolically in the air. The experiments were neither thought experiments nor crude demonstration experiments *a posteriori*; they were quantitative and were very carefully carried out. Drake and MacLachlan comment as follows:[30]

> . . . it is apparent that Galileo was describing as a mental conception something he had carefully observed with his own eyes 30 years earlier. The first historians of science jumped to the conclusion that

that was what he had done. Recent historians of science, critical of their predecessors, have jumped instead to the conclusion that Galileo worked from pure mathematics without empirical evidence; faith in ideal Platonic forms rather than attention to physical detail, they say, opened the way to modern science. As far as Galileo is concerned, the earlier historians came closer to the truth.

Another frequently aired argument against the authenticity of Galileo's inclining plane experiments goes as follows: with the technology available then (primitive clocks) it was impossible to achieve a degree of experimental precision that would permit the conclusions drawn by Galileo; hence the experiments cannot have been of any significance in the discovery of the kinematic laws.[31] This objection, however, is only valid if it is assumed that Galileo's measuring technique was comparable to that of later versions of the inclining plane experiment. According to Drake's reconstruction, this is not the case.[32] Galileo used an ingenious method based on his knowledge of theory of music and in this way was able to achieve an amazingly high level of experimental accuracy. So, following Drake, Galileo's precise data is not the result of manipulation on the basis of theory, but of an experimental technique so original and simple that it has outstripped the imagination of the historians. On the other hand, experimental reconstructions have not yielded unambiguous answers and the question cannot be said to have been definitely settled.[33] We shall return to the problems connected with experimental reconstruction in chapter 14.

The most famous of all Galileo's experiments are unquestionably the experiments that Galileo is said to have carried out with freely falling bodies from the leaning tower of Pisa. Did Galileo really carry out this experiment or is it just 'a literary legend', a myth, as many historians believe?[34] It is obvious that this alleged experiment has had an important mythical function as empiricist propaganda. But if the event took place, it is not in itself mythical. In Galileo's works, from *de motu* (1591) to *Discorsi* (1638), there are several references to free fall experiments he carried out from towers, experiments the results of which were in conflict with Aristotelian views. But the only direct evidence of the experiment from the tower in Pisa is to be found in the first biography of Galileo, written by V. Viviani (1622–1703). Viviani was Galileo's assistant during his final years and in his biography he recalled conversations with Galileo in which Galileo spoke about his experiments in Pisa. According to Viviani, Galileo demonstrated[35]

... through experience and sound demonstrations and arguments
... that speeds of unequal weights of the same material, moving
through the same medium, did not at all preserve the ratio of their
heaviness assigned to them by Aristotle, but rather, these all moved
with equal speeds, he showing this by repeated experiments made
from the height of the Leaning Tower of Pisa in the presence of
other professors and all their students.

There are, of course, grounds for being sceptical about the reliability of this account, cp. Thomson's and Henry's memoirs of Dalton.
When Galileo told Viviani about the Pisa event (if he did) it was
50 years after it had happened (if it did). Why didn't Galileo
describe the experiment himself? If it took place in the presence
of large numbers of professors and students, why have none of
these commented on it?

On the other hand: the experiment described by Viviani is in
agreement with Galileo's results in *de motu*, in which the law
about the velocity not being dependent on weight is only thought
to apply to bodies of the same material; it was only later that
Galileo realized that the velocity does not depend on the kind of
material either. If Viviani's aim was to glorify Galileo, or if Galileo
rationalized the events of his youth in his old age, one would expect
an account of experiments that showed the same velocity for all
bodies, irrespective of weight and type. And furthermore, we know
that free fall experiments of the Pisa type were carried out by
others at about the same time and with the same purpose. For
example, in 1586 at the very latest (i.e. before Galileo), Stevin and
De Groot carried out free fall experiments in Holland with lead
balls of varying weights from a height of 10 metres. It would have
been natural – in harmony with the spirit of the age and with
Galileo's personality – if he had carried out the experiment in Pisa.

Pro et contra. We do not know that Galileo made the Pisa experiment; but neither do we know that he did not. A dismissal of it
as a myth without historical reality can only be based on guesswork.
Unlike Dijksterhuis, Hall, Truesdell and others, Drake believes
that Galileo did carry out the experiment in Pisa and that Viviani's
account of this should be regarded as reliable.[36]

If the story of the Pisa experiment can nevertheless be described
as a myth, it is because the event has been distorted; in this capacity
it has achieved a legitimating function for empiricist views of science, typically in textbooks and in popular accounts. A typical

example of the mythical version of the Pisa event is due to the
eminent British physicist Oliver Lodge (1851–1940). In a book
dating from 1893 he writes as follows:[37]

> So one morning, before the assembled University, he ascended the
> famous leaning tower, taking with him a 100 lb. shot and a 1 lb.
> shot. He balanced them on the edge of the tower, and let them drop
> together. Together they fell, and together they struck the ground.
> The simultaneous clang of those two weights sounded the death-
> knell of the old system of philosophy, and heralded the birth of a
> new. But was the change sudden? Were his opponents convinced?
> Not a jot. Though they had seen with their eyes, and heard with
> their ears, the full light of heaven shining upon them, they went
> back muttering and discontented to their musty old volumes and
> their garrets, there to invent occult reasons for denying the validity
> of the observation, and for referring it to some unknown disturbing
> cause. . . . Yet they had received a shock: as by a breadth of fresh
> salt breeze and a dash of spray in their faces, they had been awakened
> out of their comfortable lethargy. They felt the approach of a new
> era. Yes, it was a shock; and they hated the young Galileo for giving
> it them – hated him with the sullen hatred of men who fight for a
> lost and dying cause.

Lodge's imaginative account has no basis in fact, of course. The
experiment carried out by Galileo in Pisa was a qualitative demonstration experiment that merely showed what Galileo already knew;
and it was no more a crucial experiment than were the earlier
experiments of Stevin and De Groot.

The debate about Galileo's method and the role of the experiment
in his science will continue. This is the type of question that cannot
be decided by merely analysing sources and that does not seem to
have any final answer. But the long and learned debate does at
least seem to have established that Galileo was neither an apostle
of empirico–inductive science nor an unequivocal hypothetico–
deductive thinker; such images of the great Italian have no basis
in historical reality, but are the result of the scientific ideals of
later ages. Moreover, most Galileo studies have assumed that
Galileo had a clearly defined methodology and worked in accordance with this; that his attitude towards the role of experiments
was unambiguous and consistent. As in other cases, this assumption
is not well grounded and appears to a certain extent to be based
on the myth of coherence (cf. chapter 9). Some of the lack of clarity
surrounding Galileo's method is undoubtedly due to the fact that
Galileo *was* unclear. The historian will have to accept this fact.[38]

13

Scientists' histories

We have seen, as the examples involving Dalton and Galileo demonstrate, that scientists' own statements cannot really be accepted as the truth without further investigation. Such statements ought to be critically evaluated and approached with scepticism. In principle this applies to all evidence, even the most primary and direct sources such as diaries, private notes, oral statements and laboratory journals. We can never be absolutely certain that the scientist who makes notes in his diary while in the middle of making a discovery really did think and behave in the way he describes. The fact that it is always possible to raise doubts about the authenticity of any source is, however, a purely negative conclusion. In practice the historian has to accept some sources as trustworthy and is justified in doing so; namely, if no other source contradicts the information given in the source in question and if, furthermore, there are no reasonable grounds to doubt its authenticity. The evidence must then be accepted as reliable. At least until anything happens to affect this status, the source can become part of the fund of historical knowledge that acts as a check on the reliability of other sources.

In the present chapter we shall discuss the value and reliability of sources written by scientists who were themselves involved in the research on which the source throws light. It is already obvious from earlier examples (cp. Dalton) that the scientist is not always a witness to the truth when it comes to his own actions. General forgetfulness and a tendency to rationalize after the event in the light of later developments will naturally play a part in retrospective accounts of events that took place many years ago. Also the scientist can have had grounds for presenting his actions differently from the way they actually were. In connection with conflicts of priority, for example, he might consciously or unconsciously overestimate his own contribution, change dates or in some other way suppress a reality that he might have wished were otherwise. It is not difficult to find examples of unreliable statements of the types mentioned. When the unreliability can be established it is because the statements cannot be reconciled with other well-documented occurrences, or because the same scientist has given conflicting accounts of the same occurrence. Dalton's account of the origin of the atomic theory provides us with an example.

Let us look at another case that has been thoroughly investigated in particular by Gerald Holton.[1] In 1887 the American physicist Albert Michelson performed a famous experiment in order to measure the velocity of the earth relative to the world-ether. Michelson's experiment was later explained by the theory of relativity and has traditionally been regarded as an experimental precondition for the establishment of Einstein's theory. The historical problem is now whether the experiment led Einstein to his theory of relativity of 1905, or in any other way played a significant part in Einstein's context of discovery, as the textbook version has it. Naturally, Einstein's own statements constitute the most direct evidence for answering the question.

We will only consider some of Einstein's retrospective pronouncements on this question, not the more indirect evidence that can be drawn from his scientific works and general thoughts from around 1900–1905. Einstein had several opportunities to say what he thought on the subject.[2] In 1931 he made a speech at a meeting in Pasadena, USA, in the presence of many American physicists and astronomers, including the 79-year-old guest of honour, Michelson, whom he now met for the first and last time. On this occasion, Einstein apparently credited Michelson with having provided the experimental basis for the theory of relativity. According to Bernard Jaffe's report of the speech, Einstein said:[3]

> You, my honored Dr Michelson, began with this work when I was only a youngster, hardly three feet high. It was you who led the physicists into new paths, and through your marvelous experimental work paved the way for the development of the Theory of Relativity. You uncovered an insidious defect in the ether theory of light, as it then existed, and stimulated the ideas of H.A.Lorentz and FitzGerald, out of which the Special Theory of Relativity developed. Without your work this theory would today be scarcely more than an interesting speculation.

Those present unquestionably regarded Einstein's words as an authoritative confirmation of what everybody knew: that Michelson's experiment had provided the basis for the creation of the theory of relativity. This is also how Jaffe interprets it: 'In 1931, just before the death of Michelson, Einstein publicly attributed his theory to the experiment of Michelson.'[4] Three points of historiographical interest ought to be mentioned in connection with the evidence from 1931 and Jaffe's conclusion.

(1) Jaffe's conclusion is a rather free interpretation of the speech, for Einstein does not, after all, express himself in quite that way. Einstein does not state that the theory of relativity had its origin in the experiment of Michelson. If this was what Einstein wished to say why did he not say it in straightforward language?

(2) The interpretation of the speech is based on what looks like a garbled quotation and on what is at the very least a dubious use of sources. For Jaffe has left out a sentence that is of importance in this context, without as much as indicating its omission. In Einstein's original manuscript of the speech (in the English translation) he said that Michelson[5]

> uncovered an insidious defect in the ether theory of light, as it then existed, and stimulated the ideas of H.A.Lorentz and FitzGerald, out of which the special theory of relativity developed. These in turn led the way to the general theory of relativity, and to the theory of gravitation. Without your work this theory . . .

'This theory', therefore, does not refer to the special theory of relativity, as one might think from Jaffe's version, but to the later general theory (1915), and thus makes Jaffe's interpretation even more dubious.

(3) Nevertheless, Einstein praised Michelson's work and drew attention to his significance for the theories of relativity in tones that at least indicate that there might have been the genetic connection that Jaffe and the standard version say there is. The reason why Einstein did not clearly hail Michelson's experiment as the experimental basis of the special theory of relativity is that Einstein knew that there was no genetic connection. But in that case why did Einstein not clearly draw attention to this instead of making a speech that hardly could fail to be misinterpreted? One must remember that the speech was made in a particular social context to a particular audience. It is an important tenet in criticism of sources that texts cannot be analysed correctly as though they were

isolated systems. Texts are always directed at a public and to a certain extent they will reflect the wishes or anticipated reactions of that public. In the case of Einstein, the atmosphere surrounding the speech was markedly empiricist; the (mythical) link between Einstein and Michelson had been pointed out by Millikan and Michelson in speeches before that of Einstein. As Holton writes, 'the stage and the expectations were fully set for Einstein's response'. In view of these expectations and the whole atmosphere of the assembly, Einstein could hardly use the occasion to publicly destroy the myth on which much of the fame of the ageing Michelson rested.

Many years later, however, Einstein made it clear that Michelson's experiment played almost no part in the discovery of the theory of relativity. R.S.Shankland carried out a series of interviews with Einstein in which the subject was touched on. From a conversation from 1950 Shankland reports:[6]

> When I asked him how he had learned of the Michelson–Morley experiment, he told me that he had become aware of it through the writings of H.A.Lorentz, but *only after 1905* had it come to his attention!

And again from a conversation in 1952:[7]

> Einstein said that in the years 1905–1909, he thought a great deal about Michelson's result, in his discussions with Lorentz and others in his thinkings about general relativity. He then realized (so he told me) that he had also been conscious of Michelson's results before 1905 partly through his reading of the papers of Lorentz and more because he had simply assumed this result of Michelson to be true.

Finally, in a letter from 1954, Einstein wrote:[8]

> In my own development Michelson's result has not had a considerable influence. I even do not remember if I knew of it at all when I wrote my first paper on the subject (1905). The explanation is that I was, for general reasons, firmly convinced that there does not exist absolute motion and my problem was only how this could be reconciled with our knowledge of electrodynamics. One can therefore understand why in my personal struggle Michelson's experiment played no role or at least no decisive role.

The conclusion to be drawn from these statements seems to be that Michelson's experiment played no decisive role for Einstein. Admittedly, statements which, like those of Einstein, deal with events that happened 40–50 years previously ought to be critically evaluated. In fact, we have reason to believe that Einstein's claim that he did not know about the Michelson experiment until 1905 is wrong. In an address given in 1922, Einstein said that[9]

> While I was . . . in my student years, I came to know the strange result of Michelson's experiment. Soon I came to the conclusion that our idea about the motion of the earth with respect to the ether is incorrect, if we admit Michelson's null result as a fact. This was the first path which led me to the special theory of relativity.

Once again we realize that scientists' own statements, taken as a whole, may be bewildering and inconsistent. To reach an unambiguous answer only by relying on Einstein's own words seems not possible.

We shall now turn to another example. The French physicist André Marie Ampère (1775–1836) is one of the founders of electrodynamics. His main work, published in 1827, is entitled *Mémoire sur la théorie mathématique des phénomenes électrodynamiques*. In this work Ampère presented his theory as having been entirely deduced from experiment, cf. the continuation of the title . . . *uniquement déduite de l'expérience*. Ampère emphasized his empirical–inductive method and its close similarity with the inductivist rules of natural philosophy established by the great Newton:[10]

> I have consulted only experience in order to establish the laws of these phenomena, and I have deduced from them the formula which can only represent the forces to which they are due; I have made no investigation about the cause itself assignable to these forces, well convinced that any investigation of this kind should be preceded simply by experimental knowledge of the laws and of the determination, deduced solely from these laws, of the value of the elementary force.

A critical reading of Ampère's work reveals, however, that this method was *not* the basis for (all) his theoretical conclusions. To Ampère the empiricist method was an ideal and a methodological creed, not his authentic practice. Ampère wished his public to regard *théorie mathématique* as being based on the highly regarded Newtonian method, and perhaps convinced himself that he followed this method. But as Duhem pointed out, Ampère's data is totally unsuited to the kind of inductive conclusions to which he refers; many of the experiments described by Ampère are imprecise and lack necessary details. According to Duhem, Ampère's authentic practice was in marked contrast to his presentation of it:[11]

> his [Ampère's] fundamental formula of electrodynamics was found quite completely by a sort of divination, . . . his experiments were thought up by him as afterthoughts and quite purposefully combined so that he might be able to expound according to the Newtonian method a theory that he had constructed by a series of postulates. . . . Very far from its being the case that Ampère's electrodynamic theory was *entirely deduced from experiment*, experiment played a very feeble role in its formulation.

Bearing in mind the largely unjustified dismissal of the experiments described by Galileo as mere hypotheses, one ought to be cautious about concluding that Ampère did not carry out the experiments he mentions. In this case, however, there is no doubt. Ampère himself admitted, though at the end of his work and almost *en passant*, that some of the experiments that he had described had not been carried out. 'I think I ought to remark in finishing this memoir that I have not yet had the time to construct the instruments represented in Diagram 4 of the first plate and in Diagram 20 of the second plate. The experiments for which they were intended have not yet been done.'[12] Ampère was obviously confident about the outcome of the experiments; so confident that it did not make much difference whether they were actually carried out or not.

Even though Ampère's text is unreliable as a representation of his real method, it is, nevertheless, an authentic expression of what he thought and of what he wanted his colleagues to connect with the work. In other cases the statements of a scientist are unreliable because they express thoughts that are not his own or that he would have expressed differently in other circumstances. The active elements in the development of science are the thoughts and actions that are expressed publicly. No matter whether the scientist means what he writes, the thoughts that have been expressed publicly will have a life of their own in the history of science. If one wishes to establish what the scientist really thought, however, and why he might have expressed a different opinion, one sometimes has to go behind the statements that were expressed publicly. Like anybody else, scientists can have many reasons for saying something that they do not think. Their real opinions might be politically unacceptable, in conflict with general morality or in embarrassing disagreement with prevailing scientific views. In such cases as these, the scientist will be inclined to modify his views, accommodate himself to the system on whose acceptance his career and penetration depend in the final analysis.

It is well known, for example, that Copernicus' *de revolutionibus* was written with one eye on the views of the Church, and that in its finished version it was marked by the attempts of the Lutheran theologian Andreas Osiander (1498–1552) to render the work harmless. Osiander provided Copernicus' work with an anonymous foreword in which the theory was presented as a pure hypothesis, not as a realistic candidate of a new cosmology. Osiander's notorious preface did not fit what Copernicus thought but it is on the face of it presented as though it had been written by Copernicus.

History of contemporary science differs from other kinds of history of science in certain areas as far as sources are concerned. Unlike all other historians, the historian of contemporary society does not have to limit himself to searching for and reinterpreting sources already in existence. He is able to create his own source materials by arranging interviews, questionnaires, etc. This fact gives certain possibilities that do not exist in history of the past. But these possibilities do not, of course, provide a universal key to the history of present science. Interviews and questionnaires, after all, only give answers to the questions that the historian thinks interesting and therefore thinks of asking; the historian's control of the situation invites perhaps manipulation of the sources.

But if the historian of contemporary science has several unique possibilities as far as sources are concerned, there are also disadvantages compared with traditional historical sources. As we have stressed throughout, non-public primary sources are of particular value because of their immediacy and authenticity. In modern scientific communities letters, diaries, informal notes, etc. are used less and less and are seldom kept. Informal contacts between scientists often take place on the telephone or in conversation at the innumerable conferences that easier communications have made possible; increasingly in ways that do not leave any permanent written record behind them that the historian can use as source materials.

It is not least in connection with sociological studies that history of recent science has been pursued. S.W.Woolgar, who has studied the discovery of pulsars, has given an illuminating account of the methodological problems in such studies.[13] As in other historical investigations, the historian (or sociologist) will first attempt to work out a chronology of the events that took place; or a 'working account' as Woolgar calls it. During this preliminary stage the historian particularly relies on review articles and broad accounts in which the discovery is discussed by specialists. He will select what he believes to be a representative section of published sources and will attach particular importance to articles that (1) were written by scientists who were actively involved in the discovery, (2) were written at a time close to that of the discovery, and (3) give detailed accounts. Usually, though, the historian will discover that these sources do not give information that tallies and cannot thus directly tell us 'what actually happened'. Woolgar's experiences of the historiography of the discovery of pulsars have a general application. He writes:[14]

> ... using several accounts of the discovery, I found it difficult immediately to discern any straightforward sequence of events. At several points the accounts did not appear to 'fit'. . . . it seemed possible that these apparent discrepancies resulted from the difficulty experienced by participants in reconstructing 'exactly what had happened'. If so, I wondered to what extent such authors would tend to order, and subsequently re-present, their recollections in what appeared to them, retrospectively, to be a 'logical' sequence. I was also aware of the likelihood of my own inclination to logical reconstruction: I might be tempted to resolve differences between two competing versions by favouring the sequence which appeared to be the more 'logical' in line with my own, unspecified, presumption.

As we have touched on earlier, this is quite a common problem in historical research: how is one to evaluate conflicting sources? As in the case of Woolgar, the historian of contemporary science can address himself directly to the scientists involved and confront them with the different versions. This can lead to a clarification of the true course of events but in many cases (Woolgar's included) will not achieve consensus. Members of a research team who were equally centrally placed in a discovery can still have very different ideas of what happened.

Instead of regarding these variations as distortions of the 'true' course of events, and attempting to establish this, Woolgar suggests that one should recognize the differences as an irreducible fact (in methodological terms). In other words, the historian can profit from using the conflicting sources as information about the process of discovery precisely because it is conflicting information.[15]

> We can regard these differences not as sources of possible 'inaccuracy' or 'distortion', but as potentially fruitful forms of data in themselves. Perhaps the very difference between the two accounts can tell us about the essential process of the development of ideas.

This way of using conflicting source materials is in line with the idea of discoveries as processes full of conflict instead of as neutral events. We have seen a similar view used by Thackray in his study of Dalton's atomic theory. It ought to be noted that Woolgar's suggestion is not merely an emergency solution in those cases in which conflicting sources cannot be reconciled. Even in those cases where the historian is able to reject or modify accounts and arrive at a picture of the authentic course of events, the fact that scientists give different accounts of it will in itself be a valuable source for understanding the process of discovery.

14

Experimental history of science

In spite of the fact that the past cannot be revoked, in one sense at least it can be investigated by experimental methods. Experimental history of science has not been used extensively or systematically and there are divided opinions about it. On the one hand, one might mention the Italian historian L.Belloni who has developed the experimental method in the history of medicine and biology in particular. According to Belloni, the reconstruction of historical experiments is of special value as a supplementary method for the interpretation of texts:[1]

> When we set to work on the study and reconstruction of the thought of an author, the analysis of his writings obviously cannot be undertaken apart from the general framework of the culture of his time. If then observations and experiments are described which in arrangement and technique are as distant from our habits and mentality as the cultural climate in which the author lived, the best and sometimes the only way of arriving at an exact interpretation of the text being considered lies in repeating the experiments under the same conditions under which they were originally performed.

On the other hand, there are historians who reject the experimental method on principle:[2]

> If we were to discover Dalton's thought, we could, it may be supposed, help ourselves by performing once again Dalton's experiments, which would put us in the situation in which he found himself thinking. We have only to assert this about Dalton to realise how unsound it is, . . . The activity of John Dalton shows us quite clearly the uniqueness of scientific acts of thought. I cannot repeat yesterday's experiment. It has gone into a past which is only accessible to the kind of enquiry we call historical.

As we shall see, these two points of view do not have to be contradictory; each one is true, in its own way.

Past experiments can be studied by the help of a (modern) experimental reproduction of them. The reason why experimental reproduction must be regarded as an acceptable method is as follows: a very significant part of the sciences has always consisted of experiments or similar empirical work. By experimenting, one manipulates natural objects in different ways and measures, or perhaps merely registers, different effects of the controlled manipulation. The effects are clearly linked to the specific experimental situation; namely, by ways of natural laws that ensure that when one arranges an experiment in a certain way, the outcome will always be determined. But since the laws of nature are independent of time, or 'ahistorical' as it were, the link between the experimental situation and the objective outcome will be valid across historical periods. We can use *our* knowledge of the laws of nature to overthrow historical reports; not about the thoughts and actions of human beings (which form the actual substance of history), but about the phenomena that, objectively, the thoughts and actions were concerned with. For example, if a chemist in the 15th century reports that he has made experiments in which he made gold out of materials that do not contain gold, we *know* that this report is erroneous. Our knowledge of chemistry and atomic theory provides us with this knowledge. The fact that the alchemists were mistaken certainly does not make their works less interesting historically. We know that it was not gold that the chemists made; but what made them believe that it was, then? If the alchemist's report is sufficiently detailed and understandable the historian can repeat his experiment today and analyse the product using modern methods. If the reconstructed experiment is an exact reproduction of the original, one can be certain of obtaining the same product as the one obtained 500 years ago. In this way one has obtained knowledge about a historical question by means of experiment. The knowledge that one can obtain about the past in this way is only possible because, in the final analysis, the ideas that are dealt with in history of science are ideas about objective features in nature. Other forms of history do not possess this qualification.

The suggested modern reproduction of historical events only applies to events that can be isolated and repeated, i.e. that are governed by causal laws. In the case of the alchemist it is just as important to know what the reasons were for the status of alchemy in the 15th century, how alchemists thought, what connections there were between alchemy and astrology, etc. The experimental method cannot be of any use here. We cannot recreate the social and religious conditions of the 15th century with any degree of certainty, cannot make them the object of actual experiments.

The historical reconstruction has a different status to the logical or *rational reconstruction* that is discussed in theory of science. In the rational reconstruction one re-thinks the problems from a particular norm of rationality and, perhaps, one criticizes a scientist for having argued in a way that is not rational. This kind of reconstruction can be valuable philosophically but as history of science it is unacceptable; it is irrelevant whether a scientist thought, or did not think, as a modern philosopher might wish him to have done. We cannot acquire any knowledge about the past by judging its events on the basis of modern norms of rationality. These norms are themselves, after all, a result of a social and historical process. As Dijksterhuis pointed out:[3]

> ... the principle, all too frequently neglected in the history of science, that if a proposition B ... really follows from a proposition A ... , a person who is acquainted with B is not on that account to be credited with the knowledge of A and the logical relation between A and B.

The experimental reconstruction can, however, contain elements of rational reconstruction. If a scientist performed a particular experiment with a particular purpose in view, the experiment can be criticized for not being rational in its context; the way in which the experiment was carried out might not, perhaps, be connected with its purpose. Such criticism can be justified but it is anachronistic as long as it does not reflect actual historical views but is due to later doctrines on the experimental method.

Repetition of experiments can only involve the physical actions that form the raw kernel of experiments: setting up of equipment, reading instruments, registering observations. This, however, is hardly an 'experiment' in the real sense. The real experiment is an integral whole in which theoretical expectations and interpretation of data are also involved. It is only in an abstract, and hence unhistorical, sense that 'purely experimental conditions' can be isolated from the theoretical framework in which the experiment is situated. In this sense one can say, as Greenaway does, that historical experiments are unique, non-repeatable occurrences. Even if, today, we repeat Lavoisier's famous experiment from 1777, in which he demonstrated the composition of air, it will not be *Lavoisier's* experiment that we are repeating. In isolation, this repetition will not be anything more than an ordinary chemical experiment. But if we make a thorough study of the scientific and intellectual climate that existed at the time of Lavoisier, the experiment can help us to understand Lavoisier better. It has become a kind of repetition of a historical experiment.

Experimental history of science can give information about whether experiments that have been reported were actually performed or whether they were merely thought. If historical texts describe results of experiments that sharply conflict with modern repetitions, one will have reason to doubt whether the experiment was actually performed in reality and gave the results described. If, on the other hand, the experiment described corresponds to reconstructed experiments we have reason to believe in the authenticity of the report. The modern control that we can exercise in this way does not have to be experimental. It will often merely be a theoretical check that the experimental results that have been described agree with modern established knowledge. Thus we do not need to repeat the experiments of the alchemists in order to know that they did not make gold.

Two qualifying remarks may be necessary here: with reference to the 'verification' mentioned above the mere fact that the historical report agrees with modern knowledge is not in itself sufficient grounds for accepting the report. If X, in 1750, describes an experiment in which, after passing through a prism, white light is seen to be divided up into a colour spectrum, the fact that the experiment is in agreement with modern knowledge should not lead to an acceptance of its historical authenticity. Such experiments were well known in 1750. We have to demand that the outcome of the experiment is new or surprising for the time. If, in 1650, Y describes the same observation there are good grounds for believing Y. For such results as this were not known and not theoretically predictable in 1650. How could Y have reported a correct result of a non-trivial experiment, if he had not performed the experiment?

With reference to the 'falsification' of historical accounts, one cannot just conclude that if the reported result conflicts with modern knowledge or with the result of the reconstructed experiment, then the experiment was not really performed as described. Then, as now, people who make experiments report what they regard as its important features, on the basis of their theoretical expectations and of the purpose and context of the experiment. A scientist may well have observed a phenomenon that we know to be true and that the reconstruction upholds, without having reported it or having been in any way aware of it. He might have regarded it as irrelevant 'noise', while the historian of science, repeating his experiment with the advantage of hindsight, will regard the phenomenon as interesting and meaningful. In *Discorsi* Galileo reported his experimental discovery that pendulums oscillate isochronously, i.e. that the period of oscillation does not depend on the amplitude (how far the pendulum moves from its position of rest). According to Galileo he observed that the period was exactly the same for each amplitude. This is not, in fact, the case with wide amplitudes and Galileo just cannot have measured it in his experiments. Even so, there is no reason to doubt that Galileo's reports are authentic. Galileo knew that the period is not quite constant with wide amplitudes; but he regarded the deviation as unimportant and hence reported it as non-existent.

It is not only that the circumstances surrounding the experimental report makes simple falsification as good as impossible. Actual testings will also require a comprehensive knowledge of the earlier experimental situation and method so what is repeated is the exact historical experiment. Such knowledge is often not present and the reconstructed experiment will contain such high levels of uncertainty in relation to the original that it is not possible to draw any conclusions.

An example can again be found in *Discorsi*. In this work Galileo (Salviati) refers to a strange experiment with water and wine.[4] A glass globe with a little hole is filled with water and placed, with the hole facing downwards, against a bowl of red wine. I now saw, says Galileo, that the red wine went up into the globe while the water went down into the bowl, without the fluids becoming mixed; at the end, the globe was full of red wine, the bowl full of water. Did Galileo make that experiment, whose result may seem to be in conflict with what we know about the motion of fluids? Koyré evidently accepted the applicability of the experimental method, for he writes:[5]

> It is, indeed, difficult to put forward an explanation of the astonishing experiment he [Galileo] has just reported; particularly, because, if we repeated it *exactly as described*, we should see the wine rise

in the glass globe (filled with water) and water fall into the vessel (full of wine); but we should not see the water and the wine simply replacing each other; we should see the formation of a mixture.

This historically reconstructed experiment is itself fictitious. Koyré does not claim to have performed the experiment, but thinks he knows what will happen. Koyré concluded that 'Galileo . . . had never made the experiment; but, having heard of it, reconstructed it in his imagination, accepting the complete and essential incompatibility of water with wine as an indubitable fact'.[6] The Canadian historian of science James MacLachlan, however, went to the trouble of actually repeating the experiments.[7] He was able to support the result reported by Galileo. Conclusion: Galileo's experiment with wine and water is presumably authentic.

The history of chemistry, marked as it is to such a high degree by experimental work, ought to be especially amenable to the experimental historical method. H.C.Ørsted (1777–1851) and Friedrich Wöhler (1800–1882) share the honour of having discovered aluminium. Ørsted has priority since he reported his reduction of alumina (aluminium oxide) to metal in 1825, while Wöhler's improved method stems from 1827. In earlier works on the history of chemistry, Ørsted's discovery was often ignored so that it was felt necessary to determine whether Ørsted had actually isolated aluminium in 1825. In 1920, on the occasion of the centenary of Ørsted's discovery of electromagnetism, Danish chemists repeated the procedure described by Ørsted and in this way extracted pure aluminium.[8] There can thus be no justifiable doubt that Ørsted did actually make aluminium in 1825 (whether he also discovered the element is a slightly different question).

The fact that there can be any doubt at all about Ørsted's method is partly bound up with Ørsted's original report which cannot lead to aluminium, if one follows it word-for-word. For Ørsted writes:[9]

Dry chlorine was conducted over a mixture of pure alumina, which was kept red hot in a porcelain tube. Since the alumina was thus able to separate from its oxygen, its combustible parts combined with the chlorine and in doing so formed a volatile compound that was easily caught in a receiving flask that naturally had to be provided with a waste pipe for the unabsorbed chlorine and the carbonic oxide gas [carbon monoxide] produced. The compound of the chlorine and the alumina's combustible element, the aluminium chloride, is volatile at a heat that does not much exceed that of boiling water; it is slightly yellowish, though perhaps from attached carbon, it is soft but takes the form of a crystal; it absorbs water greedily, and easily dissolves in this and with the development of heat.

It is obvious that the reported experiment cannot give the results described since the metallic part in alumina is aluminium oxide. Where could the 'carbonic oxide gas produced' and the 'attached carbon' come from if carbon were not also present? Aluminium oxide does not in fact react directly with chlorine. The words 'carbon black' (coal dust) have been omitted, presumably because of a printing error, so the correct sentence would read 'over a mixture of pure alumina and carbon black, which . . . '. This, at least, was the direction on which the confirmation of Ørsted's experiment rested in 1920. With this revision Ørsted's report not only becomes chemically intelligible but also linguistically consistent. The expression 'a mixture of pure alumina' does not make sense if it is not a mixture with something; in view of the later reference to carbon monoxide this something must be interpreted as being carbon.

But, one might object, with what right can the historian revise an original text and add words that do not actually appear in it? Is it not conceivable that Ørsted might have made the experiment exactly as described and therefore *not* have prepared aluminium? In Ørsted's case might this well-meaning revision of his text not be an example of 'the mythology of coherence' if the only grounds for it were the above-mentioned considerations of common sense? No, not in this case. The hypothesis about a trivial printing error receives documentary support, for in the same year Ørsted described his discovery in a letter to the German chemist Schweigger, where he writes: ' . . . One obtains the aluminium chloride as a volatile substance when one conducts dry chlorine over red-hot alumina mixed with carbon . . . '.[10] The conclusion, therefore, is that there actually was an unfortunate printing error in Ørsted's original report and that Ørsted did prepare aluminium in 1825.

When one uses modern scientific methods for the study of the past one draws elements into the past that were unknown to it. This does not mean, however, that it is anachronistic historiography. It is a purely technical intervention into the past without any later knowledge being attributed to the people of the past. There is nothing illegitimate about this kind of intervention.

All historical insight has its starting point in modern analyses. The historian uses his current situation and knowledge, including the results of modern science. There is nothing new in this or anything peculiar to history of science. Chemical analyses of paper and ink and physical methods for dating records have long been used in historical and archaeological research. Modern knowledge used as an instrument in history and archaeology has been systematically developed in modern times and is known as archaeometry. These techniques have most relevance for general history of civilization and in particular for its pre-literate periods where the usual analysis of sources cannot give us much information.

A recent analysis of Newton's hair provides us with a curious example of archaeometry relevant to history of science. In his youth, Newton went through a period of severe neuroses and religious mania, and Newton scholars have given the most divergent explanations of these. The fact that the remains of Newton's hair reveal an abnormally high concentration of mercury and the fact that mercury is known to cause mental injury provide us with a new explanation. A much more simple and prosaic explanation, but possibly rather more trustworthy than the complicated psychological guesswork of historians.[11]

The relevance that history has for the present can sometimes take on a completely concrete character in history of science. A lot of scientific work only exists in historical archives in an unprocessed form. One might imagine that this data might be of some value to the modern scientist, if the historian of science was to make it accessible. In practice, however, one must regard such a 'historically' based science as Utopian and, therefore, agree with C.A.Elliott in the following judgement:[12]

> It does seem very unlikely that any scientist would come to a conventional archive to consult, in conjunction with his own current work, that data of an earlier scientist. Except in a formal sense, he probably does not even consult the very early published documentation.

Nevertheless, historical data can be of value to modern science in some cases, possibly even the only source for the answering of questions. The natural sciences, including astronomy, geology and evolutionary biology, include aspects which are not repeatable and are, in this sense, 'historical' sciences. The palaeozoologist studies the physical relics of extinct animals. If even the relics have disappeared he is forced to work as an historian. A focal point in modern geophysics is the study of how the mass of the earth is distributed around its centre and the reason for this. The study of early Babylonian cuneiform texts have proved to be of great importance for this question: the Babylonians reported careful observations for the moon and the planets, including lunar eclipses which they timed in relation to sunrise or sunset. If modern scientists calculate at what time of day an eclipse, visible in ancient Babylon, happened on a particular date, it turns out that there is a discrepancy between the Babylonian records and the calculated value. The discrepancy is not reasoned in inaccurate Babylonian data but in the fact that the length of the day is gradually increasing because of the slowing down of the earth's rotation. Comparison of modern calculations with Babylonian records indicates that the change in the speed of the earth's rotation cannot be due only to tidal effects but must also depend on changes in the distribution of mass inside the earth. The non-tidal effects can be estimated rather accurately on the basis of the cuneiform texts which thus help geophysicists improve their models of the interior of the earth.[13] The Babylonian data has of course to be interpreted and made intelligible by the historian of science before it can be communicated to the geophysicist. Also in other areas of the history of astronomy there are examples of how historical data has been successfully used in modern research and of how it could have played a useful part if it had been utilized.[14]

However interesting these examples of the relevance to science of historical data might be, it should not be forgotten that they are exceptions; and they can hardly be called proper history of science.

15

The biographical approach

Biographies of eminent, individual scientists are one of the oldest forms of history of science. However, in the new, professional history of science it has been regarded as a less-esteemed form of history, to some degree. It is only recently that this trend has been reversed.[1] The diminishing respectability of the biography is connected with modern standards of scholarship in history of science and with a change in general perspective where the focus has to some extent moved to either intellectual or social topics. Biographical works are still, however, an important part of history of science and they will remain so. Even though biographies are often of dubious quality, as seen from a history of science point of view, they can carry out functions not covered by other forms of history.

Since the scientific biography is built up around the activities of an individual it can easily veer towards giving a distorted picture of the development of science. Namely by, in the very nature of things, concentrating on the achievements of the scientist whose life story is being told, and thereby possibly glorifying these, while other scientists merely appear as a grey background. The fact that a biography is written from a person-centred perspective does not in itself merit criticism and is not in itself a sign of lack of objectivity. The biographer, however, will often be tempted to identify himself with the subject and present the portrayed scientist as a hero; while his opponents and rivals are presented as villains. When this happens, the biography degenerates into so-called hagiography, uncritical black and white history. There is no doubt at all that scientific biography sets the stage to a tempting degree for the kind of black and white painting that Agassi describes as inductivist history of science.[2] It is no coincidence that myths of anticipation and other forms of mythical history of science flourish in the biographical literature.

The mythicization of history that is a common feature in so many biographies is connected with the fact that the biography is often directed at a wider public. Biographies are almost the only kind of history of science literature that succeeds in becoming best-sellers. But these widely read works, such as Eve Curie's biography of her mother, *Madame Curie*, rarely live up to the standards one would like to associate with the scientific biography. If a biography is to attract wide present-day interest it will have to appeal to the reader either by setting the stage for modern disciplinary connections or through its contents of human drama. If these elements do not exist in the real life of the biographical subject, the biographer is tempted to invent them or imagine them. Who can be bothered to read a whole book about a scientist who admittedly played an important part in his day, but whose contribution has been revealed to have been a blind alley and whose life was, furthermore, undramatic?

The biography that glorifies and romanticizes will typically present the hero as a genius struggling against a stupid contemporary world that placed every kind of obstacle in the way of his brilliant ideas; ideas that are brilliant because they anticipated, or can be read into, modern knowledge. Such obstacles will often not have any authentic basis in fact but will merely be a means of strengthening our admiration for the hero (if he overcomes them) or of excusing his lack of success (if, in spite of everything, he does not overcome them). As we have already seen, this kind of myth is not confined to the more popular type of history of science. It is obviously the duty of the historian to puncture myths, where these can be located.

As to romance, few cases in the history of science can equal the tale of the death of the French mathematician Évariste Galois (1811–1832). According to the standard story, propagated by virtually all of Galois's biographers, Galois was a misunderstood genius whose brilliant theories were suppressed by the mathematical establishment. A victim of circumstances, he became involved in the political turmoil of the time and imprisoned because of republican sympathies. Even in the prison did Galois continue to develop his mathematical ideas, later known as group theory. In 1832 the young Galois became involved in an unhappy love affair which, according to the standard story, resulted in a duel of honour with a political enemy. The night before the duel Galois 'spent the fleeting hours feverishly dashing off his scientific last will and testament. .... What he wrote down in those last desperate hours before the dawn will keep generations of mathematicians busy for hundreds of years'.[3] Galois, only 20 years old, was killed in the duel.

Unfortunately for romantics, the story is largely a myth. Recent scholarship has argued that Galois was not an innocent victim of circumstances but rather a republican hothead with an almost paranoid antipathy against authority. As to the duel, it seems to have been a result of a personal quarrel and neither caused by love affairs nor politics. Galois's alleged 'scientific last will and testament' is a legend: the night before the duel Galois was indeed occupied with mathematics but actually of a rather trivial sort, viz. making editorial corrections on manuscripts. The destruction of the Galois myth yields a more authentic history without diminishing the scientific originality of Galois. If it makes his biography a little less exciting, this is a cost which should not be regretted. T.Rothman, who has contributed to the undermining of the Galois myth, writes:[4]

> His [Galois's] reputation is not served, however, nor is the history of science, by a legend that insists a scientific genius must be above reproach in his personal life, or that any contemporary who does not appreciate his genius is either a fool, an assassin or a prostitute. The notion that genius is not tolerated by mediocrity is too old a platitude to be adopted uncritically as accurate history.

Even though mythicization and black and white painting are frequently appearing elements in biographies, there are, after all, many examples of scientific biographies that are anything but hero worshipping. Newton has been a favourite subject for hagiographical works ever since his death, usually portrayed as a majestic genius completely absorbed in his science. In Frank Manuel's scholarly biography of Newton, the great physicist is pictured as a genius indeed, but a human genius who suffered psychical conflicts to the edge of paranoia and who was in no way above worldly concerns. Manuel's portrait of Newton is scarcely flattering but it is a truer, better documented and much more interesting portrait than that produced by earlier Newton hagiographers. By the same token, Westfall's Newton 'was a man like all of us, facing similar moral choices in terms not altered by his intellectual achievement. .... Newton's role in history was intellectual not moral leadership'.[5]

The biographical approach to the history of science can be accused of giving a narrow, individualized and internalist picture of the development of science; of focusing on the individual genius at the expense of the collective and social currents. It is true, of course, that biography focuses rather unambiguously on the individual plane and that history of science would give a grossly misleading picture of history if it only consisted of biographies. But in the first place biography is only one solitary instrument in the history of science orchestra. And in the second place the focusing of biography on the individual will not necessarily happen at the expense of collective and social factors. In fact biography, in one version, can be decidedly externalist; it can, for example, depict the subject of the biography as a mere medium for social and economic currents typical of the time. In such cases the real protagonist of the biography is not the person but the super-individual currents of which he is seen as an exponent or a medium. Now it is a fact, not disturbed by the debate about the motive powers of history, that science is primarily created by single individuals. No matter what influence external factors have on scientific development, it is individual human beings who throughout history have thought the thoughts and performed the experiments that are the backbone of science. The social and institutional frameworks only become effective when they are mediated through individual living people. So, at its best, biography is the 'literary lens', to use Hankins's expression, through which we can study the impact of the external factors on science.

One of the big advantages of the biographical method is that it permits an integrated perspective on science. If one wants to have a true picture of how the philosophical, political, social and literary currents of a period interact with science, one can profitably focus on the individual. The individual is a unit through which these currents pass through the same 'filter', are mixed, and come out as science in some form or other. But, naturally, we cannot expect that the process unveiled in this way will be typical. One of the obvious limitations of biography is that it does not make generalization possible. In his defence of history of science biography, Hankins expresses the advantage mentioned above as follows:[6]

> We can say at least one thing with certainty about biography: the ideas and opinions expressed by our subject came from a single mind and are integrated to the extent that the person was able to

integrate them in his own thoughts. We have, in the case of an individual, his scientific, philosophical, social and political ideas wrapped up in a single package. This package will most likely contain contradictions, blind spots, and irrelevancies. Often the individual will appear to keep two provinces of his mind completely distinct (usually at the precise point where we are looking for a connection), but such is the perversity of human kind, and we are more honest if we accept this perversity rather than try to extrapolate the rise of science onto a smoothly ascending curve.

It is precisely the integrated perspective that is difficult to fulfil in practice. It is tempting to divide a biography up into two separate sections, especially when the science in question is difficult to understand or not clearly bound up with extra-scientific events in the person's life. The course of the scientist's life is then described in the first part, his science in the second part. Such a division is rather common, presumably because it makes the biography available to a larger circle of readers: the reader with specialist knowledge, who wants to concentrate on the science can do so without being disturbed by 'irrelevant' biographical details; the non-specialist reader can get the full benefit of the first part and skip the second part. The disadvantage is, of course, that any connection between the subject's scientific and extra-scientific activities hereby disappear. Admittedly, an opposite danger can be found in integrated biography, the tendency towards exaggerated integration; for example, to always see the scientific contributions of the subject as being based on or related to extra-scientific events. The artificial integration can be just as misleading as the artificial isolation.

Biographies naturally include aspects of the portrayed scientist's psychology. If psychoanalytical or similar ideas are used extensively one may talk of a *psychobiographical* approach. This is a difficult art, full of pitfalls.

Sigmund Freud, the father of psychoanalysis, wrote a psychobiographical study of Leonardo Da Vinci in which he analysed experiences from Leonardo's childhood. In this study Freud made a blunder in mistranslating the Italian word for 'kite' with 'vulture', a term which has a symbolic significance in psychoanalysis and on which Freud based parts of his interpretation of Leonardo.[7] A modern example of Freudian psychobiography is Lewis Feuer's study of the psychological roots of modern physics, based on the idea that scientists' mental activities express a sublimation of emotional tensions.[8] For example, Feuer explains Ernst Mach's opposition to atomism by referring to young Mach's strained relations with his authoritarian father. What has father-hatred to do with atoms? Feuer argues as follows: Mach sometimes referred to atoms as 'stones', which is the biblical metaphor for testicles. Since Mach hated his father, he dreamt of a de-pater-nalized reality, that is, a reality without 'stones'. By association Mach projected this psychological revolt into a physical theory in which there was no need for atoms.

In this case, like in Freud's, terms are conceived symbolically and arbitrarily subjected to psychological interpretation without further evidence. A less artificial explanation of Mach's use of the term 'stone' would be that Mach thought of the smallest building stones (Bausteine) of matter, not of testicles. But then, of course, there would be no need to psychologize. The example may illustrate the danger in applying pre-conceived, psychological notions to biographical events. Only if the events seem otherwise inexplicable, that is, cannot be explained on rational grounds, psychological data or psychoanalytical reasoning should be considered. In the case of Mach psychologisms hide the fact that anti-atomism was far from unusual at the end of the 19th century and that Mach's attitude was in fact based on good, scientific reasons.

Whether psychologically orientated or not, biographies of indi-viduals can only play a limited part in history of science. The biography is situated inside a framework that by its very nature it cannot transgress. Thus, in time it will be confined to the gener-ation to which the portrayed person belongs, and it will similarly be geographically confined. Furthermore, in practice, the biography will only be concerned with a special type of scientist: the great scientists whose works have been of pioneering importance, who have been influenced by philosophical ideas and who have possibly also played a public role – the aristocrats of science. The thousands of less important or less exciting scientists will remain beyond the reach of biography. If one wants to capture the typical scientific environment of a particular period, not just its élite, one can hardly do so by means of the individual biography.

16

Prosopography

The historical technique based on collective biographies and similar sources is called prosopography. What characterizes this method is that it uses data concerning many people and events as its sources.

Prosopography is not a method that is peculiar to history of science and it is only recently, in fact, that it has been introduced into this field in an elaborate form. This happened via inspiration from general social history, especially from economic history, that has long used quantitative methods of the same type as those used in prosopography.[1] However, collective biographies have been used sporadically in the study of science for over 100 years, starting with Francis Galton (1822–1911) who compiled statistics about eminent British scientists so as to study the relationship between heredity, environment and genius.[2] The statistical studies on geniuses made by Galton and others were strongly influenced by the extreme social-Darwinism of the Victorian age; today they are regarded as classic examples of so-called scientism. Wilhelm Ostwald used membership of scientific academies as a measure of 'greatness' and studied the distribution of members of such institutions with regard to sex, race, religion and nationality.[3] Among other things, Ostwald concluded that women had no scientific ability and that Teutonic men had a particular aptitude for science (it is perhaps unnecessary to state that Ostwald was a German male). Studies of ability and genius like those of Ostwald and Galton are not, of course, *comme il faut* today, although their research methods in more refined forms have been taken over by modern quantitative sociology and historiography. Galton, Ostwald and other early scientists in the same tradition are clear illustrations of the fact that quantitative historiography is not a particular objective method and can easily degenerate into ideology.

If anyone can be called the precursor of modern prosopography, it is not Galton but the Swiss botanist Alphonse de Candolle (1806–1893). In 1873 he wrote an ambitious work, *Histoire des sciences et des savants depuis deux siècles*, in which he systematically used statistical methods for the study of the factors that promote or hamper scientific progress. Candolle investigated the dependence of science on hereditary and institutional factors by relating the careers of leading scientists to their educational background or to the careers of their parents. In many ways, Candolle's work staked out the perspectives and methods that were to form the basis many years later for modern sociology of science and science of science. Although Candolle's book was translated into German, at Ostwald's instigation, in 1911, it did not have any immediate influence on history of science.[4] It was not until the middle of the 1930s that the sociologists P.Sorokin and R.K.Merton used similar methods in the history of science and technology (see the next chapter). Merton was inspired by Candolle's book which, though neglected by historians of science, was well known to sociologists. The *Handbuch zur Geschichte der Naturwissenschaften und Technik* published by Ludwig Darmstaedter in 1908 played an important part in Merton's investigations as it did in other investigations in the same tradition. The book is a chronologically ordered collection of approximately 13,000 discoveries and inventions.

The sources used by the prosopographically orientated historian differ from the typical sources of intellectual history of science. Analyses of the contents of scientific publications, of letters and manuscripts, are not particularly interesting to the prosopographer. The sources suited to his purpose are collective biographies, tables of discoveries, protocols and yearbooks of scientific institutions, academic registers and many other things. A first step will often be to consult biographical dictionaries like the *Dictionary of National Biography* (England). For 19th century science in particular, Poggendorf's old *Handwörterbuch* is an unrivalled source of biographical data that has played a role similar to that of Darmstaedter's *Handbuch* in prosopography.[5]

Studies of the development and careers of scientific communities and disciplines are a genre that utilizes methods that are similar to those of prosopography. In this genre, the interest lies in how a particular scientific discipline originates, is developed and disintegrates, for example; what the social structure of the discipline is like; its paradigmatic basis; which people are members of the community, and how they are related to each other; how the characteristics and values of the discipline are transmitted to new geographical areas and specialist fields. In recent years many case studies of this kind have appeared, including experimental psychology, mathematics, molecular biology and radioastronomy.[6]

Thus, Mullins investigated how the study of phages (a type of virus) developed from around 1935 when a few scientists (in particular Max Delbrück) established a research programme that soon evolved into a vital scientific discipline. Mullins is not interested in the progress of knowledge in microbiology but in the social processes in the development of the new discipline, such as hierarchy, recruitment, communication and status. Information about pupil/teacher relationships and forms of co-operation is supplied by collective biographies and secondary literature. In the period from 1945 to 1953, when phage research took the form of a separate discipline, Mullins summarizes its structure in a network that shows the connections between the few scientists then involved in the discipline (Figure 2). A chart like this gives a description of the structure of a scientific community during a particular period but says nothing, of course, about the science in question as regards

Figure 2. The network of phage research, 1945–1953. Reproduced from Mullins (1972), p. 60, with the permission of Minerva.

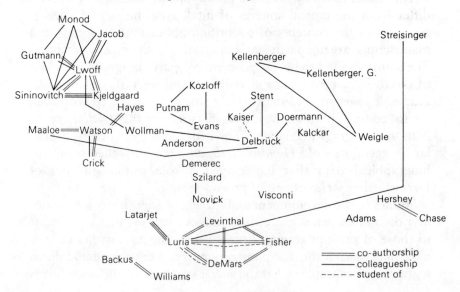

contents. Furthermore, this type of network can easily be over-interpreted and, without justification, be accredited great objectivity. The 38 scientists who appear in the chart must have been chosen by Mullins on the basis of certain criteria or ideas about what phage research actually was during the period. If other criteria had been used, the members of the discipline would not be the same ones and the network would look different.[7]

The development of a scientific discipline is partly determined by how effectively it is 'marketed'. In the transmission of ideas, personal contacts between scientists, the recruitment of new disciplines, and the creation of social structures to which members feel themselves committed, are important elements. Many modern historical and sociological studies have concluded that it is these factors, and not primarily the truth of scientific ideas, that determine the growth of science. In Mullins' study of phage research, the successful development of the discipline is seen as the result of effective marketing with production of students, the setting up of courses and the establishment of professional standards. Fisher investigated the development of another, less successful scientific discipline, the mathematical theory of invariants. This theory constituted a progressive research area in the last quarter of the 19th century but then degenerated and eventually died out; by about 1930 invariant theory no longer attracted the interest of mathematicians.[8] Fisher's study of the fate of invariant theory illustrates how a scientific discipline can die out if one does not ensure that it is carried on through the recruitment of students. Science does not live on by itself, on the strength of its intellectual qualities alone.[9]

As has already been intimated, there is a temptation in prosopography, as in all forms of quantitative historiography, to regard one's data as unproblematic and purely empirical. 'Prosopographers have sometimes tended to present themselves as objective, experimental scientists who do not need to consider historiography since they feel they have revealed the numerical essence of an historical problem.'[10] It should be obvious that this is unjustified. 'Phage research' and 'invariant theory' do not have any clearly defined or natural demarcation lines, but are the result of an interpretation or an estimate and in this sense are non-objective. The prosopographer cannot, just as other historians cannot, avoid qualitative historiographical considerations.

The social history that collective biographies lead up to is not the history of ordinary scientists, but rather the history of the aristocracy of science. Most prosopographical studies have concentrated on the élite, from the members of the Royal Society in the 17th century to the exclusive circle of Nobel Prize winners from the present century. In one sense this is only natural, since collective biographies usually only cover the élite. Even such broad reference works as *World Who's Who in Science* (containing 30,000 names) only contain professional practitioners of science. As have been emphasized by Pyenson, Thackray and Shapin, there are good grounds for widening the social history of science so as to embrace not only the run-of-the-mill scientist who never received any professional recognition, but also the many non-scientists from the periphery of science. Pyenson writes:[11]

> In the history of science these figures might be reflected in the legions of popularizers who explicated the conservation of energy or the theories of relativity, the journalists and essayists who created the climate for receiving scientific ideas and financing scientific

Figure 3. The figure to the left shows the cumulative number of authors in phage research, the one to the right the cumulative number of publications in invariant theory. Note that a cumulative linear growth, as in the invariant theory 1895–1915, corresponds to stagnation, that is, the same number of new articles per year. When the graph is almost horizontal as from 1935 to 1941 this means that almost nothing is published. Reproduced from Crane (1972), p. 177 and p. 178, with the permission of Chicago University Press.

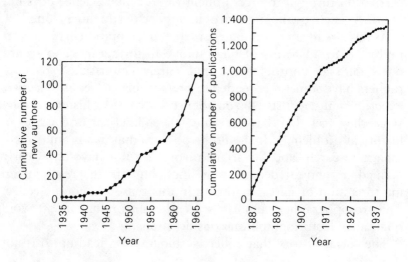

projects, the scientific publishers and editors who earned their living by catering to public tastes, and the university trained scientists who published little 'scientific' work beyond their dissertations. 'Common ideas' in science might be clarified by studying the lives and work of such populations, rather than those associated with the institutions and organizations that sheltered ruling élites in science.

At the present time there are only few studies in existence in the broad social history sense that Pyenson suggests. We shall mention two examples.

Arnold Thackray is a central figure in modern prosopography. He investigated the scientific milieu in England in the 19th century as it developed around the many privately organized societies. One of the most important of these societies was the *Manchester Literary and Philosophical Society* (MLPS), whose most famous member was John Dalton. What part was played by the MLPS at the time? Why was the society created, which people were members, and what did they expect from it? The answering of such questions as these invites the use of prosopographical methods, for the significance of such an institution as the MLPS cannot be evaluated solely on the basis of scientific publications. Thackray studied biographical data concerning about 600 early members of the MLPS.[12] It appears that the overwhelming majority of these came from the growing middle class (doctors, industrialists, merchants, bankers, engineers) and very few of them had much knowledge of or specialist interest in science. Thackray's data leads him to the conclusion that the real function of the MLPS was not to pursue or promote science but to give social legitimation to the interests and cohesion of the new class. One did not gather around science for the sake of science but for the sake of the ideology. According to Thackray science became 'the predominant mode of cultural expression' for the members of the Manchester Society; 'natural knowledge was the private cultural property of a closely knit, continuing intermarrying, almost dynastic élite'.[13]

In another study, together with Steven Shapin, Thackray considered the social role of science during the industrial revolution in England.[14] Thackray and Shapin argue that this role will be incomprehensible as long as one identifies science with cognitive interests and expert knowledge. Only by including under the label 'science' (*natural knowledge* rather than *science*) all the people who have in some way or other been involved in scientific circles, will it be possible to understand the social role of science. Thackray and Shapin therefore call for a radical change in the history of science perspective:[15]

> Historiographically, we have been accustomed to disregard science as it percolates from men of science to the generally literate. It has either been dismissed as non-science, scientism (hence, irrelevant or pernicious), misunderstood science (hence, error), or popularized science (hence, trivial). In point of fact, *science as people think of it and as they use it* is every bit as historically important as science as scientists conceive of it. . . . Recent work hints that we have grossly underestimated the extent to which scientific ideas, scientific principles, attitudes and modes of enquiry permeated the social structure and served important structural and dynamic purposes in the process of British industrialization. As we begin to look at natural knowledge in the varied cultural contexts in which it has flourished, we shall probably come to see that perspectives carried from the university laboratory are far from adequate to business of history. In creating a deeper historical understanding of the scientific enterprise, prosopography is by no means an all-sufficient tool. It is nonetheless a highly promising, and as yet insufficiently exploited, mode of conceptualization.

Shapin's works on the Edinburgh science culture in the early 19th century is an example of the new prosopographical historiography. Shapin deals in particular with the development of phrenology.[16]

According to the doctrines of phrenology, the brain was the organ of the mind and the site of a number of distinct psychological faculties which could be read off by examining the contours of the skull. Phrenology gained wide popularity in Edinburgh in the 1820s, receiving considerable support by the lower middle classes; it was, on the other hand, rejected by the upper classes and the university establishment. The attempts to disseminate phrenology led to a protracted controversy the core of which was as much a matter of social power as of scientific truth. In such a case the audience – the people who were either sympathetic or unsympathetic to the cause of phrenology but did not actively participate in the debate – is just as important as the actors and can profitably be treated prosopographically. For example, in order to establish the social affiliations of the two camps Shapin makes use of collective biographies of the Fellows of the Royal Society of Edinburgh (anti-phrenological) and the members of the local Phrenological

Society. In this way he is able to show that phrenology was an outsider movement, attracting many merchants, artisans, engineers and lawyers but virtually no university professors.

The fact that in the cases investigated by Thackray and Shapin it has proved fruitful to operate with a 'prosopographical view of science' does not, of course, ensure that such an approach has general validity. The Manchester Literary and Philosophical Society and the phrenology movement have not very much in common with mainstream physics, for example. A very substantial part of scientific development, in particular in modern time, has taken place virtually without an audience. In those cases the prosopographical model of science has little relevance.

17

Scientometric historiography

The term scientometry is used here to denote a collection of methods for analysing structure and development in science at a relatively highly developed level. As a methodological discipline, scientometry does not have any special object field; the methods can be applied in other social forms of organization than scientific ones, without any significant changes being necessary. Scientometry is not a particularly historical technique either although it is in that capacity that we will be concerned with it here. In fact, in many respects it is linked up with present science and must rather be called a quantitative sociology of science technique that can also be applied to parts of earlier science. In integrated studies of science – science of science – scientometry plays an important role as an instrument of analysis and prognosis for research policy.

One can distinguish between two kinds of studies within scientometrically orientated history of science, namely:

1. Studies that focus on the temporal development of science, quantified in various ways. Typically the development of scientific growth.
2. Studies that focus on the structure of scientific communication in a given period or on the influence of scientific contributions in the period. This form of history of science is close to many prosopographical and sociological studies.

Truly quantitative studies of history of science are a new phenomenon. The first completely quantitative history of science study – scientometrical in the sense we are using here – is from 1917, in which Cole and Eames applied bibliometrical methods to the history of anatomy.[1] Another early example is the Soviet scientist Rainoff, who studied the development of physics on the basis of statistical analysis of the literature, number of discoveries, etc.[2] In this way Rainoff was attempting among other things to correlate the fluctuations in scientific development on the one hand and social and economic history on the other. Rainoff's work had no influence at the time but contained much of the foundation on which scientometrists built decades later. Merton's important work from 1938, *Science, Technology and Society in Seventeenth Century England*, in which extended use was made of quantitative analysis of source materials, was of more importance for the later development. One of the ways in which Merton illuminated the controversial question of the connection between the science of the time and the politico–economic conditions was to analyse the topics dealt with at meetings of the Royal Society and other forms of scientific communication. By classifying and counting these topics he found the percentage of research that was driven by socio–economic needs, and used this data for his conclusions about the relationship between science and economics.[3]

Before looking at further, more modern, examples of quantitative historiography, we will give a critical résumé of the methods and basic assumptions of scientometrics.[4]

Whenever science is measured quantitatively, the chosen limitation of what science is becomes decisive and problematic. For a measure to be suitable for the quantity called science one has to demand

- that it is reasonably 'realistic' in the sense that it agrees with the general, qualitatively based view of what science is.
- that it is reasonably 'objective', that is, the chosen measure should not be ambiguous or subject to idiosyncracies.

However, these two demands are usually not reconcilable. If, for example, one wants to judge which are the most progressive fields of a scientific discipline at a given time, one can use interviews with leading scientists (informants), yearbooks, state-of-the-art reports, etc. Such a technique as this will undoubtedly be able to deliver a realistic, relevant picture of the state of the science in question; but it will be a rather 'subjective' picture that cannot easily be quantified. Alternatively, one can investigate the question bibliometrically, by counting how many publications are written in the fields of the science in question and how these have varied over time. This is a very 'objective' method, but on the other hand it will only superficially, and perhaps not at all, reflect the conditions that one actually wishes to study.[5]

Scientometrics makes use of two quantitative measures in particular in order to illuminate the growth and distribution of science, the number of scientists and the number of scientific publications. In the first case one will typically use the number of scientists, or the number of scientists per 100,000 inhabitants, for example, as a gross indicator of the level of scientific activity. But even at this stage there are obvious problems: which people should be classified as scientists? according to which criteria? One might, for example, use one of the following criteria:

(a) Scientists are the people who are employed in institutes and similar places, whose purpose it is to pursue science.
(b) Scientists are the people whose names appear in scientific bibliographies, surveys, standard reference books and collective biographies.
(c) Scientists are the people who have published articles or books on scientific topics.

There are obvious weaknesses in each of these criteria, especially when it comes to historical application. Thus criterion (a) will exclude all the amateur scientists who have notoriously played an important part in science up to this century. Whatever criterion of the term 'scientist' is used, it will merely put off an operational definition of the term 'science' itself. To search for such a definition which is, at the same time, historically meaningful, seems futile.

The crucial point is that any quantification of science presupposes an understanding of the nature of science. It is important to insist on this, in particular because quantifications often appear as and are presented as being based on objective, unproblematic criteria; which they never are, no matter what measure is used.

It is mostly in connection with science before our century that there will be problems with scientometrical standard measures. Shapin and Thackray have argued that (in the case of England at least) science must be seen as a social institution in which it is difficult to differentiate contributors from receivers, scientists from non-scientists. When science is seen as 'an ecologically well-adapted variant which answered the social and ideological needs of many whose participation in literate culture might otherwise have been non-existent', the whole basis of the so-called objective measure of the scientometrists crumbles.[6] According to Shapin and Thackray many of those who published science in the period 1700–1900 were not scientists, not even when a very liberal definition is used. They did not contribute new knowledge. Only a minority of the members of the scientific societies did any research at all, and even fewer of them published anything in the periodicals of the societies.

The most frequently used measure in scientometrics is the direct output of scientific activity, namely publications. This measure has been developed in particular by Derek de Solla Price, one of the founders of scientometrics and modern quantitative history of science.[7] According to Solla Price scientific activity differs uniquely from all other cultural and social activities by fundamentally being *universal, objective, cumulative* and *papyrocentric*. By the last term is meant that the output of science is paper of some kind or other (books, articles, preprints, pamphlets). Solla Price has defined science as 'that which is published in scientific papers', and a scientist as 'a man who sometime in his life has helped in the writing of such a paper'.[8] It is obvious that such a definition of science is operational with respect to quantitative studies, but it is just as obvious that it is open to criticism.

The definition presupposes a papyrocentric scientific community in which publication is a recognized virtue or a necessity. There is no doubt that basic science today is ruled by a 'publish or perish' phenomenon; but in earlier times the tyranny of publication was less explicit or did not exist at all. It will therefore be misleading, in the case of much early science, to equate science and publications. Furthermore, the definition assumes an unproblematic distinction between 'scientific papers' and 'non-scientific papers', or between periodicals that contain science and those that do not. With respect to early science in particular, it is not possible to transfer the rather precise meaning of our own scientific specialist periodicals without absurdity. In earlier centuries much excellent science was published outside these channels, often in very obscure contexts.

Solla Price and others have formulated a so-called exponential law that is supposed to reflect the fact that the 'size' of science grows exponentially.[9] As an example of the exponential law, look at Figure 4. This curve apparently shows that at least as far as size is concerned physics has displayed a steady exponential growth throughout the period. Large-scale external events (the two World Wars) even reveal themselves as temporary declines that do not, however, have any influence on the rate of growth over longer periods. Solla Price used this curve to give his message concerning the frequently asked question about the influence of war on science. Does war stimulate science or does war have an inhibiting effect? Solla Price's answer:[10]

The graph shows immediately that neither of these things happened – or, rather, if they did, they balanced each other so effectively that no resultant effect is to be found. Once science had recovered from the war, the curve settled down to exactly the same slope and rate of progress that it had before.

Such a conclusion, however, is only possible if by science one means the quantity of publications within all physical science disciplines combined. In fact there is no reason why one should interpret the seductively regular growth in the curves of Solla Price and others as a corresponding growth in scientific knowledge or use them as support for a cumulative conception of science. Thus it is a distinctive feature of the chart that it does not show any periods of boom or periods of stagnation. The revolutionary ideas that physics produced in this century cannot be traced at all in the figure. This is a natural consequence of the fact that interesting or pioneering works do not count for any more in statistics of publi-

Figure 4. The total number of *Physics Abstracts* in the period from 1900 to 1955. Reproduced from Solla Price (1974), p. 172, with the permission of Yale University Press.

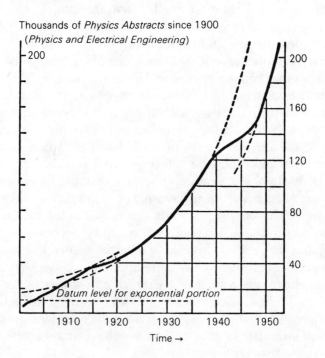

Thousands of *Physics Abstracts* since 1900
(*Physics and Electrical Engineering*)

Datum level for exponential portion

Time →

cation than the great mass of mediocre or indifferent works. It is only because cumulation and continuity are built into the metric employed that it can lead to the absurd result that there have not been any particularly dynamic phases in 20th century physics.

Part of the problem with primitive publication statistics is, of course, that they measure something different to what the historian and theorist of science are really interested in, the development and quality of scientific knowledge, the conceptual innovations, and so on. The relationship between productivity and 'quality' in individual scientists has been studied by several analysts. This has mainly been done by replacing the subjective, not very operational, term 'quality' with relatively objective terms like 'success' and 'professional recognition', the latter to be measured as membership of prestigious scientific societies or as scientific honours received.[11] On the basis of such methods Solla Price maintained that 'on the whole there is . . . a reasonably good correlation between the eminence of a scientist and his productivity of papers'.[12] Whether there is such a correlation or not, 'eminence' or 'success' is not the same as quality. Many of the most original and innovative scientists in history have not had 'success' in the world of science; only a few of the most productive authors can be counted among those normally regarded as important contributors to scientific progress.[13]

An alternative way of quantifying science to statistics on publications is to rely on a counting of 'scientific achievements', important discoveries or events. These events can, for example, be counted from chronologies or surveys. This was Rainoff's method in the previously mentioned work from 1929. Rainoff based his data on discoveries in Auerbach's *Geschichtstafeln der Physik* and allocated the same value to each of the discoveries counted ('for the purpose of accounting', as Rainoff wrote). Similar methods have often been used later, though seldom with anything but unreasonable results. For example, one can plot the number of known elements cumulatively against time, as shown in Figure 5.[14] Dobrov regards this curve as an illustrative example of the course of development of chemistry and more generally of the 'alternation between periods of stormy developments and periods of a certain languor' that is characteristic of all scientific development and that is linked up with 'the dialectic process of the transition from a quantitative accumulation of new facts, experiences, methods, etc. to qualitative changes in the content of science itself'.[15] It is especially hazardous in this case, however, to use the curve for anything other than an amusing overview. It is anachronistic in the sense that it only encompasses elements that have achieved accepted status according to modern knowledge; it unashamedly operates with the modern concept of elements in periods when the view of what constitutes an element was quite different. The few hundred elements that turned out to be mistakes or that were based on an older view of elements are not included in the curve. In the kind of teleological history that the curve represents, there is no room for such abberations as Lavoisier's 'calorique', Mendeleev's 'newtonium' and Lockyer's 'protometals'; abberations that were, nevertheless, discoveries.

Other studies use the number of scientists recorded in standard biographies, the number and distribution of scientific awards, and the number of discoveries recorded in chronological surveys as their measure of scientific activity.[16] A recent example is Simonton's

Figure 5. The number of recognized discoveries of chemical elements as a function of date. Reproduced from Solla Price (1963), p. 29, with the permission of Columbia University Press.

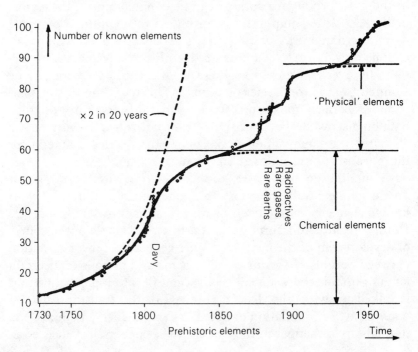

attempt to explain the relationship between war and scientific creativity.[17] Simonton takes his measure of 'creativity' or 'scientific discovery' from a table that gives about 10,000 'important scientific discoveries and inventions' distributed between times and countries.[18] However, a measure like this must be regarded as unacceptable: the number of those discoveries that Darmstaedter regarded as important enough, in 1908, to be included in his work cannot reasonably be a satisfactory measure of scientific creativity; even less so since neither Darmstaedter, Sorokin or Simonton make any attempt to differentiate between the importance of these discoveries. In the case of Simonton, as in others in the same tradition, the tables and surveys that act as data banks predetermine, to a high degree, the conclusions. Not even the most advanced statistical methods can change this.

Solla Price has used a similar method in an ambitious attempt to objectively determine times of boom and times of stagnation in the development of science, and thereby give a firmer basis to periodization.[19] It might be worth quoting Solla Price's working hypothesis, which is that 'the objectivity and transnational character of basic science lend to its historical development a much larger element of determinism and of imperviousness to local socio–economic factors than one is accustomed to elsewhere in human affairs'.[20] This view of science, which is characteristic of the scientometric tradition, leads to a particular historiography of science strategy:[21]

> It follows that a vital task of the historian of science and technology is to analyze such quasi-automatic secular change as proceeds regardless of particular causes, for only then can we dissect out those non-automatic and significant events that require special *ad hoc* explanation. We need to perceive and understand regularity of behavior before we can get to a second-order explanation of the deviations therefrom.

Solla Price has then counted all the scientific events in a number of chronological works of the Darmstaedter type, and has worked out the deviations from the regular, exponential growth rate. The result appears in Figure 6 which, in the opinion of Solla Price, reveals a valuable, objective division into periods. The reason why Solla Price has so much faith in the curve is that it corresponds well with the historian's intuition about the periods of boom and stagnation.[22]

The fact that in reality one checks the result via comparison with qualitative historical insight, implies that one accepts this insight, after all, as a kind of check list. This applies generally to quantitative historiography and thereby places the assertion about a better and more objective perspective in a problematical light: if quantitative history of science really does have a superior status, why should it then be necessary to evaluate and correct its results by comparison with 'subjective' historical insight?

Scientometric techniques are based on the largely tacit assumption that at least in principle it is possible precisely to localize scientific discoveries as events in time and that the development of science can be understood by the cumulative addition of such events. This shows up clearly in the chronological technique just mentioned, when a specific discovery is allotted to a specific year. But such a view is primitive and misleading. Scientific discoveries are not usually discrete events; they are processes that can seldon be localized to a particular time or particular place. Furthermore, it can often be difficult to decide whether a discovery was actually made at all or whether it was *turned into* a discovery retrospectively.[23]

Figure 6. Variations in the scientific and technological development. Reproduced from Solla Price (1980), p. 183, with the permission of D. Reidel Publishing Company.

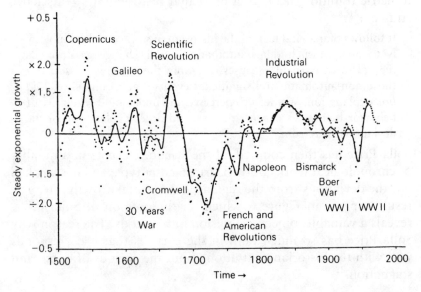

We can provisionally conclude that quantitative historiography, based on counting scientists, publications or discoveries, is encumbered with considerable methodological defects and an inbuilt bias. We will now look at another kind of quantitative historiography that has won some influence in recent years, namely techniques that make use of frequency and distribution of citations. The reason for using citation measures, i.e. measures that are defined on the basis of how many times a scientific publication is quoted by other publications, is that a measure like this is supposed to give a more realistic picture of the impact of a scientific work. Unlike publication as a measure, the citation measure has the advantage that it can be used in connection with a single scientific work. If this work is frequently quoted by colleagues in the discipline it will tend to be judged important and of 'high quality'.

If one wishes to study the frequency with which a particular work is quoted, one can simply count the references to it in all the other works that one thinks relevant. This is a time-consuming and uncertain job but it is the only possible method as far as older science is concerned. For literature after 1961 one can profitably use the *Science Citation Index* (SCI), published currently, which systematically covers about 3,000 selected periodicals as well as most books.[24] Among other things, SCI gives information about how many times a work is referred to, by whom and in which publications. Although SCI only starts with 1961, similar data banks can also be made for earlier periods.[25]

Is the citation measure a reliable measure of impact or value in the sociological sense? An answer in the affirmative presupposes that the members of the scientific community live up to the norms generally accepted today of always referring to the works one has relied on for the information; and only these works. This presupposition is problematic.

For example, the use of 'cosmetic references', i.e. references that do not have any actual importance for the work in question, is a widespread phenomenon in modern science. This is partly due to the fact that a large number of references is thought to give more weight to an article, make it look more imposing. Or it may be due to the fact that one feels it opportune, for collegiate or diplomatic reasons to refer to one's professor or peers. An author X, who has independently developed a theory and who finds out just before publication that another scientist Y has previously developed a similar theory, will normally see that a reference is made to Y; without one being able to conclude from this that X was influenced by Y.[26]

Modern works that build on classic results from early science seldom refer to these. An important part of the literature will almost systematically avoid being cited because it is included as 'tacit', assumed knowledge that everyone in the speciality concerned knows about without it being necessary to make any direct reference to it. Furthermore, conflicts of priority, or other forms of controversy, can easily result in the publications of the opponent deliberately being left out of the reference list. If a scientist wishes to promote himself at the expense of a rival, this can happen by ignoring the publications of the rival. In tense situations, wars or political crises, it can become a downright patriotic duty not to acknowledge contributions from the opposite side. This happened during the First World War and the years following it, when militant scientists in England and France recommended ignoring contributions from Germany or in the German language. The German physicist Arnold Sommerfeld commented on this deliberate ignoring of German science in a letter to Niels Bohr, whom he thanked for never having failed to make references to German physics: 'Through this, the colleagues in the same field in the enemy countries, who normally want to suppress all German achievements, will presumably also be made to realize that German science will not allow itself to be kept down, not even during a war.'[27]

Another break with the scientific ethos is the more or less overt plagiarism that has always been a real phenomenon in science.[28] In such cases the references will not cover the publications that form the basis for the plagiarism. In a study of the sociology of modern high-energy physics Gaston discovered that about 50% of the physicists interviewed believed that at some point or other they had not been cited when they ought to have been. One of the informants said: 'It very often happens that people who haven't published much will not refer to your work because the only way they can get their paper into print is by not referring to the preceding paper that has done the same thing.'[29]

The above reservations lead to the conclusion that citation measures cannot, without qualification, be accepted as reliable measures of impact. There is no doubt that in many cases frequency of citation does reflect impact, but the measure should not be credited with any particularly objective or reliable status compared with evaluations based on a qualitative assessment.

Citation networks have been used to identify the contributions that are specially important in a scientific discipline ('key papers' or 'nodal publications'); namely, as the contributions to which other publications in the discipline very frequently refer. In Figure 7 an attempt is being made to ascertain whether Mendel's famous work from 1865, in which he laid the foundations of genetics, remained unknown to his contemporaries, as the standard version

Figure 7. The citation pattern concerning Gregor Mendel's article from 1865. Reproduced from Garfield (1970), p. 670, with the permission of Macmillan Journals Limited.

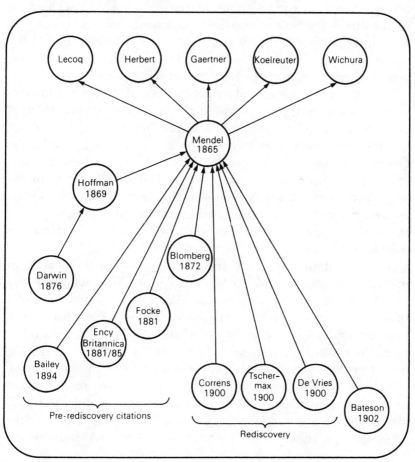

asserts in the history of biology. Mendel's discoveries were rediscovered in 1900 by de Vries and others and it was not until then that his works had any influence on the development of biology. The fact that Mendel's paper, published in 1866, was cited by five publications, including the *Encyclopedia Britannica*, between 1869 and 1894 shows that it was not totally ignored. The periodical in which Mendel's article appeared, *Verhandlungen des naturwissenschaftlichen Vereins in Brunn*, was no obscure publication either; 115 libraries and scientific institutions in Europe, including the Royal Society, subscribed to the *Verhandlungen*. In addition, Mendel sent 40 reprints of his paper to botanists and other natural scientists. On the other hand, the citation network cannot be taken as proof that Mendel's paper was, after all, well known. But it can at least give rise to some interesting questions, such as: if Mendel's work was not, apparently, unknown, why did it not play any important part? Why was its value not acknowledged? Why was it necessary to rediscover it? Had Darwin, who cites Hoffman, who in turn cites Mendel (five times at that), read Mendel?[30]

The citation technique will not, in itself, be of any help in answering these questions; partly because references do not give information about the referring author's understanding of the reference or about the context in which he refers to it. For example, the German doctor W.Focke (1834–1922) referred to Mendel's discoveries in a book that was bought by Darwin. But Darwin did not learn about Mendel's works in this way. Darwin's copy of Focke's book still exists today and there is good reason to believe that Darwin never read it;[31] the part in which Focke referred to Mendel is still uncut. With reference to H.Hoffman (1819–1891), a German botanist, we know that Darwin read the publication in which a reference is made to Mendel. Although Darwin probably saw Hoffman's short reference to Mendel, we cannot conclude that then Darwin also knew about Mendel's investigations. It is only by studying the contents of Hoffman's reference, in other words by working on qualitative lines, that one can see that 'despite the surprising proximity of Mendel and Darwin, the fateful dice of history seem to have been loaded against any intercommunication'.[32] It appears from Hoffman's reference that he did not understand, and at any rate never mentioned, the point of Mendel's investigations; the point that would have interested Darwin.

In a series of studies, Sullivan *et al.* have carefully examined the development in so-called weak interaction physics, a branch of elementary particle physics.[33] Inspired by Lakatos's theory of science, Sullivan tried to identify what is known as progressive, stagnating and degenerating phases in the development of the field from 1950 to 1972. One of the methods was to study the correlation between theory and experiment in the period by seeing how many experimental works cited theoretical works. According to Lakatos, the relationship between theory and experiment is a decisive indicator of whether a research programme is progressive or not.[34]

The result of Sullivan's studies is that theory and experiment remained separate for the whole period, since the 'citation overlap' was not statistically significant at any point. Interpreted in the Lakatosian way, this result should mean that weak interaction physics did not experience any progressive phase during the period, a conclusion that flatly contradicts the views of the physicists involved. So here again we meet the conflict between qualitative and quantitative evidence and are forced to ask which form of evidence is the most reliable. The conclusion of Sullivan et al. clearly outlines the dilemma:[35]

> . . . if we take a very strong position on the relevance of these data, and if we take Lakatos seriously, we must conclude that weak interactions subsequent to 1959 was experiencing a period of 'declining progress'. How seriously and confidently one can take Lakatos' analysis here is, of course, open to debate. At least one empirical case, that of radioastronomy as described by Edge and Mulkay, seems to be in obvious conflict with Lakatos, since it presents a picture of a rapidly growing speciality in which experiment almost always led theory. One could, of course, resolve this inconsistency simply by overriding the accumulated wisdom of those who have watched the field and who (despite the pre-eminence of experiment) claim that radio astronomy was thoroughly 'progressive' during the period of Edge and Mulkay's study: one might, for instance, just classify radio astronomy as 'stagnating', *by definition*, because of the relation between experiment and theory. But that hardly seems sensible.

One of the last shoots on the stem of scientometric techniques is the so-called *co-citation analysis* that has especially been developed by Henry Small. When an author A cites both author (article) B and author C, this double citation could be taken as an expression of the fact that in A's eyes contributions B and C are related. The intensity of co-citation is defined as the number of times a pair of publications are cited by others. If there is intensive co-citation the pair will be regarded by the practitioners of the discipline as belonging to the same group and as forming an 'intellectual focus'. Pairs of citations can similarly be related to each other, for example (C,D), (D,E), (C,F), for the purpose of identifying the 'cognitive kernel' of a speciality. In this way, by making co-citation patterns year for year, one can hope to follow the temporal variation of the intellectual foci, identify the emergence of new foci, and so on. The technique of co-citation is narrowly based on the SCI bibliographies and will therefore have only a limited application in the history of earlier science.[36]

We can now make conclusions about the scientometric approach to history of science:

In its endeavours to establish an objective historiography, scientometry focuses on a model where science is perceived as a flow of discrete information whose cognitive contents, in principle, do not matter. The atoms of the flow of information are the recognized publications such as have been canonized in SCI or other bibliographical works. Because of this, one can hardly avoid giving a too formalized picture of processes that in reality are strongly marked by informal and non-rational influences. As Edge has pointed out, scientometry gives priority to the formal rather than the informal; while traditional historiography, in contrast, tends to explain the formal on the basis of the informal.[37] It should be obvious that scientometry cannot, in any circumstances, stand alone. If it has to have any historical value it must be regarded as a supplement, and occasionally as a corrective, to traditional historical methods. This is also the view of most scientometrists, though not always their practice. If scientometry is used with care and in combination with other methods, it can play an important part, especially in the study of modern science.

Notes

All references refer to the bibliography.

Chapter 1

1. Sailor (1964), reprinted in Russell (1979), pp. 5–19. Cf. also Hunter (1981).
2. Bailly (1782), vol.3, p. 315.
3. Priestley (1775), pp. VI–VII.
4. Ibid., p. XI.
5. Leibniz (1849–1863), vol.5, p. 392.
6. Whewell (1837), vol.1, p. 42.
7. Extensive bibliographical information is given in Engelhardt (1979).
8. Translated from Ørsted (1856), p. 122.
9. Quoted from Engelhardt (1979), p. 112.
10. Steffens (1968), p. 28.
11. Whewell (1867), p. 186.
12. Liebig (1874), p. 256.
13. Darwin (1872).
14. Todhunter (1861), Todhunter (1865), Todhunter (1873).
15. Whewell (1837), Whewell (1840).
16. Mill (1843). Cf. Losee (1983).
17. Ostwald (1889).
18. Sudhoff (1910).
19. Mach (1960). For Mach's conception of history of science, see Blüh (1968) and Hiebert (1970).
20. Mach (1883), quotation from the English edition of 1960, p. 316.
21. Kopp (1843–1847).
22. Hoefer (1842–1843). Details on the historiography of chemistry is given in Weyer (1974).
23. The professor was a certain Pierre Laffitte who was a leader of the positivistic church in Paris but completely incompetent as a historian of science. See Paul (1976).
24. Translated from Fichant and Pécheux (1971), p. 52.
25. Comte (1830), as translated in Andreski (1974), p. 52.

26. Engels (1886), here quoted from the Danish translation, Marx and Engels (1971), vol.2, p. 372.
27. Schorlemmer (1879).
28. Du Bois–Reymond (1886), p. 271. Cf. Mann (1980).
29. Jagnaux (1891). See also Bensaude–Vincent (1983).
30. Draper (1875).
31. The literature dealing with the history of medicine is vast. For an introduction, see Pelling (1983).
32. Guerlac (1963), p. 807. Detailed bibliographical information on the development of history of science can be found in Thackray (1980) and Corsi and Weindling (1983).
33. Tannery (1912–1950), vol.10, p. 106. Here quoted from Hall (1969), p. 212.
34. Duhem (1905–1907), Duhem (1906–1913), Duhem (1913–1959).
35. Duhem (1905–1907), vol.1, p. 111.
36. Wohwill (1909).
37. Heiberg (1912).
38. Merz (1896–1914).
39. Dannemann (1910–1913). Darmstaedter (1906).
40. Dannemann (1906). Grimsehl (1911).
41. Duhem (1974), p. 269.
42. Sarton (1936), Sarton (1948), Sarton (1952).
43. Rupert Hall describes Sarton as 'an immensely learned man' but adds that 'one cannot but wonder, with all respect, if he was ever a historian at all'. Hall (1969), p. 215. According to Thomas Kuhn, 'historians of science owe the late Georges Sarton an immense debt for his role in establishing their profession, but the image of their speciality which he propagated continues to do much harm even though it has long since been rejected'. Kuhn (1977), p. 148.
44. Sarton (1936), p. 5.
45. Sarton (1927–1948).
46. Libby (1914). Here quoted from Thackray (1980), p. 456.

Chapter 2

1. 'The distinction between history as it happened (the course of events) and history as it is thought, the distinction between history itself and merely experienced history, must go; it is not merely false, it is meaningless.' Oakeshott (1933), p. 93. Cp. the discussion in Danto (1965), pp. 71 ff.
2. Pedersen (1975), p. 8.
3. Cf. McMullin (1970).
4. Pearce Williams (1966a).
5. Hankins (1979), p. 15. The quotation does not express Hankins's view but that of leading historians of science such as Whiteside, Truesdell and Aaboe.

The tension between scientifically and historically oriented history of science is discussed in Reingold (1981).

6. Kuhn (1977), p. 136.
7. Beyerchen (1977). Many of the most respected historians of science have never had a scientific training. It may suffice to mention the names of Edwin A. Burtt, Alexandre Koyré and Herbert Butterfield.
8. Canguilhem (1979), p. 8.
9. There are, of course, numerous examples of works which unjustifiably neglect the technical aspects of the science they deal with. Thus Lewis Feuer analyses at length the contributions of Einstein, Bohr and Heisenberg apparently without having read or understood their scientific works. See Feuer (1974).
10. Sarton's definition of history of science, as given in Sarton (1936), thus projects a modern empiricist ideal of science on to the past. As mentioned in chapter 1, Sarton's definition rendered many important men of science, such as Galen, scientifically uninteresting.
11. Ross (1962).
12. Shapin and Thackray (1974), p. 11.
13. Gillispie (1970–1980), vol.1, preface. As to the demarcation of 'science', the editors of the *Dictionary* stated as their policy to cover those areas 'that in *modern times* fall within the purview of mathematics, physics, chemistry, biology and the sciences of the earth' (my emphasis). *Ibid.*
14. As witnessed by a growing number of specialized periodicals, e.g. *Technikgeschichte, Technology and Culture* and *History and Technology*.
15. Newton's executor, Thomas Pellet, considered the stacks of non-scientific copies and manuscripts left behind to be 'foul and waste papers' which were 'not fit to be printed'. The first major Newton biographer, the physicist David Brewster, was embarrassed when he recognized Newton's unorthodox interests and accordingly minimized them in his biography. Brewster (1855).
16. Dobbs (1975). Also Westfall (1980), pp. 285 ff.
17. Boas and Hall (1958).
18. See the contributions in Bonelli and Shea (1975) and, for a recent review of the debate, Vickers (1984).
19. Figala (1977), Figala (1978).
20. Westfall (1976), p. 180.
21. Jacob (1977), pp. 99 ff.
22. Archaeoastronomy deals with prehistorical astronomy. In recent times the field has attracted a good deal of attention and is now established as a sub-field of history of science. It has its own journal, *Archaeoastronomy* (first published 1979), which exclusively deals with prehistorical astronomy. The idea that Stonehenge was designed as a sort of astronomical observatory was argued in the last century by Normal Lockyer but has only been substantiated in the last few decades. See Thom (1971). However,

the archaeoastronomical interpretation of megalithic monuments has not been accepted by all specialists. An archaeologist thus considers it as 'a kind of refined academic version of astronaut archaeology. . . . The interpretations appear to be subjective and imposed by the observer'. Daniel (1980), p. 71.

23. Childe (1964), p. 15.
24. Collingwood (1980), p. 58.
25. Forbes and Dijksterhuis (1963), vol.1, p. 11.
26. An example: In his much praised biography of Einstein, Abraham Pais fails to comment on some of Einstein's works on the following grounds: 'Since this subject [relativistic thermodynamics] remains controversial to this day, it does not lend itself as yet to historic assessment.' Pais (1982), p. 154.
27. Cf. Hendrick and Murphy (1981).
28. Giere (1973), p. 289 and p. 290.

Chapter 3

1. Truesdell (1968), p. 305. According to Truesdell some recent physicists, including himself, have been led to new results in rational mechanics through the study of Cauchy's works, dating back to 1820. Another, more spectacular example is Pieter Zeeman's Nobel Prize-rewarded discovery of the so-called Zeeman effect in 1896. Zeeman was inspired by Maxwell's account of Faraday's unsuccessful attempt to trace the influence of magnetism upon spectral lines. After studying Faraday's original work, Zeeman repeated it with more advanced experimental equipment and immediately detected the effect that Faraday had missed.
2. Jammer (1961), p. VII.
3. Schrödinger (1954), p. 16.
4. Hooykaas (1970), p. 49.
5. Conant (1961), p. 327. See also Kuhn (1984a), p. 30.
6. Mikulinsky (1975), p. 85.
7. Considering the wide circulation of Bernal's *Science in History*, it is worth noting that Bernal's views, too, belong to the same category of arguments that refer to the present. The entire message of Bernal's work is that history of science is justified because it can show the value of scientific progress and show its dependence on social conditions. Bernal (1969), in particular pp. 1219 ff.
8. For a brief bibliographical review, see Wood (1983).
9. Kröber (1978), p. 68.
10. Ibid., p. 67. See also Hahn (1975).
11. See, e.g. Brush and King (1972).
12. Woodall (1967), p. 297.
13. Whitaker (1979). Brush (1974).
14. Sarton (1948), p. 57.

15. Snow (1966).
16. Jaki (1966), p. 505.
17. Clark (1971), p. 296. A similar attitude can be found in Krafft (1976).
18. Butterfield (1949), pp. VII–VIII.
19. Cohen (1961), p. 773.
20. Pearce Williams (1966b). Various attitudes to the purity and relevance of the history of science are discussed in Elzinga (1979). Pearce Williams's purist attitude makes him conclude not only that scientists are generally incompetent as historians of science but also that philosophers should keep away from the field. Pearce Williams (1975).
21. Thackray (1980), pp. 16–20.
22. Elkana (1977), p. 257.

Chapter 4

1. Ranke (1885), p. VII. Here quoted from Marwick (1970), p. 35.
2. Quoted from Marwick (1970), p. 54.
3. Sarton (1936), p. 10.
4. See, e.g. the excellent discussion in Schaff (1977), pp. 181 ff.
5. Carr (1968), p. 30 and p. 11.
6. Hesse (1960). The impossibility of establishing absolute scientific facts is a general feature which is relevant not only to the early sciences. Historians who insist on building their accounts on such facts will run into troubles in modern science too. For the extensive literature which support this claim, see Shapin (1982).
7. Carr (1968), p. 22.
8. Beard (1935). It should be remarked that Ranke was not really a positivist historian in the sense of the nineteenth century. In fact he rejected the positivistic claim of reducing history to 'social physics' and emphasized that the sum of facts about the past is in no way identical with the history of the past.
9. The quotation is from Goethe's *Farbenlehre*, published in three volumes in 1810. Here translated from Canguilhem (1979), p. 15.
10. Greene (1982), pp. 19–68.
11. Quoted from Schaff (1977), p. 107.
12. Beard (1935), p. 75. For a precise and critical appraisal of Beard's theses, see Dray (1980), pp. 27–46.
13. Cf. Schaff (1977), pp. 194 ff.
14. Danto (1965).
15. In this context 'practical needs' is not to be understood in the sense of material needs. According to Croce, practical needs can be 'a moral requirement, the requirement of understanding one's situation in order that inspiration and action and the good life may follow upon this. It may be a merely economic requirement, that of discernment of one's own

advantage. It may be an aesthetic requirement, like that of getting clear the meaning of a word, or an allusion, or a state of mind, in order fully to grasp and enjoy a poem; or again an intellectual requirement like that of solving a scientific question by correcting and amplifying information about its terms through lack of which one had been perplexed and doubtful'. Croce (1941), p. 17.

16. Nietzsche (1874).
17. Carr (1968), p. 26. My emphasis.
18. Dewey (1949).
19. Finnochiaro (1973), pp. 202 ff.
20. Quoted after Schaff (1977), p. 95.
21. Collingwood (1980), p. 215.
22. Ibid., p. 304.
23. Ibid., p. 296.
24. Collingwood's appraisal of Einstein's science compared with that of Newton shows clearly that the idea of rational reconstruction is part of Collingwoodian re-enactment. Einstein's knowledge of Newton's theories, although based on current textbooks and not on Newton's own works, is sufficient proof to Collingwood that Einstein re-enacted Newton's thoughts. 'Newton thus lives in Einstein in the way in which any past experience lives in the mind of the historian, as a past experience known as past . . . but re-enacted here and now together with a development of itself that is partly constructive or positive and partly critical and negative.' Collingwood (1980), p. 334. For the modern notion of rational reconstruction, see Lakatos (1974).
25. Ibid., p. 308
26. Cf. Hall (1969), pp. 217–219.
27. The reflectivity argument against scepticism in its strong form may not be logically compelling, but it certainly weakens the scepticist position. See Collins and Cox (1976), p. 430.

Chapter 5

1. Bloch (1953), pp. 48–60. Nagel (1961), pp. 576–581. Atkinson (1978), pp. 42–51.
2. Popper (1969), p. 189.
3. Bloch (1953), p. 54.
4. Hooykaas (1970), p. 48.
5. Blake (1959), p. 338.
6. Popper (1961), p. 150.
7. Ibid.
8. Schaff (1977), pp. 275 ff. Cf. also Mandelbaum (1971) who provides an elaborated argument in favour of objective historiography.

9. Hermerén (1977).
10. Gilbert and Mulkay (1984), p. 124. A sound criticism of the approach of Gilbert and Mulkay is Shapin (1984).
11. Collins and Cox (1976), p. 438.

Chapter 6

1. Knight (1975), p. 32.
2. Lilley (1953), p. 58.
3. Hempel (1942), reprinted in slightly modified version in Hempel (1965), pp. 231–243. See also Gardiner (1952). Nomological knowledge is concerned with non-particular, generalizable aspects of objects, considered to be representatives of classes rather than being individual objects.
4. The example is taken from Finnochiaro (1973), p. 39.
5. Ibid., p. 52.
6. Gallie (1964), p. 108.
7. Roll-Hansen (1980), p. 513.
8. Dray (1957).
9. Adapted from Laudan (1977), p. 166.
10. Ibid., p. 188.
11. Finnochiaro (1973), pp. 53–55.
12. The requirement of explanation symmetry can be seen as related to the so-called strong programme in sociology of science. According to the strong programme explanations of scientific knowledge should be impartial and symmetrical with respect to truth and falsity, success or failure. Bloor (1976), p. 5.
13. See Shapin (1982) and Dolby (1980).
14. Nye (1981).
15. Brannigan (1981), p. 70.
16. Watkins (1959), p. 505. See also Hayek (1952).
17. Popper (1961), p. 157.
18. Watkins (1953), p. 729.

Chapter 7

1. Needham (1943), p. 12.
2. Bernal (1969), p. 1297.
3. Ibid. Notice that Bernal seems to consider explanations to be answers to why-not questions, thus agreeing with the view of Finnochiaro. Bernal's counterfactually formulated explanation of the discovery of radioactivity is surprisingly naive, being a case of the often criticized 'Cleopatra's nose historiography' (Cf. Carr (1968), p. 93). Bequerel's discovery was, according to Bernal, 'a real accident in the history of science', occasioned

only by the accidental talk with Poincaré. *Ibid.*, p. 734. (This talk, incidentally, took place in 1896, not in 1897.) However, as shown in Badash (1965), the discovery of radioactivity was far from accidental.

4. Nagel (1961), p. 589. The role of counterfactual conditionals in scientific theories is a complex subject. See, e.g. Goodman (1955), pp. 13–35 and Suppe (1977), pp. 36–45.

5. Gould (1969).

6. Kuhn (1978), p. 171.

7. Canguilhem (1979), p. 143.

Chapter 8

1. Olszewski (1964). Olszewski recommends that the historian should make use of 'the periodization [which] corresponds to the internal logic of the process under consideration'. p. 195. Thus he suggests that in the history of aeronautics Leonardo's ideas, dating at about 1500, should be regarded as later than the Montgolfier balloon flight (1783). According to Olszeweski, Leonardo's premature considerations and the gliding flights of Lilienthal (1891) are to be regarded as historically contemporaneous. For a more general defence of the use of non-chronological time in history, see Kracauer (1966).

2. Spengler (1926).

3. Sarton discusses the question of the relative importance of historical periods in Sarton (1936), pp. 20 ff.

4. Crombie (1953), p. 1.

5. Koyré (1968), p. 21. 'Modern physics', in Koyré's terminology, means the physics of the seventeenth century.

6. The concept of the scientific revolution was introduced by some French authors at the end of the 18th century. See Cohen (1976).

7. Greene (1985), p. 102.

8. Greene (1982), p. 121.

9. Cf. Agassi (1963), pp. 7 ff.

10. Dolby (1977). Various mechanisms for the transmission of science are also discussed in Crane (1972).

11. However, the often repeated story that Cannizaro's Karlsruhe address instantly convinced the chemists of the truth of Avogadro's hypothesis is a legend. Furthermore, what Cannizaro marketed as 'Avogadro's hypothesis' differed substantially from what Avogadro had suggested in 1811. See Fisher (1982) and Morselli (1984), pp. 176 ff.

12. Laudan (1977), p. 173

13. Lepenies (1977), p. 59. Lepenies distinguishes between a traditional, disciplinary 'history of science' and a new, interdisciplinary 'historical study of science'.

14. Knight (1975), p. 25.

15. Lovejoy (1976), p. 5.
16. Ibid., p. 15.
17. Sachs (1976), p. 125.
18. Sambursky (1963), pp. 96, 135 and 137. For another example, see Gunter (1971) in which the philosopher Bergson (1859–1941) is credited with having proposed the essentials of modern, relativistic cosmology years before Einstein. Other versions of the invariance thesis deal with more comprehensive categories, whole world views. Fleck (1980) developed in 1935 a theory of invariant, collective and archetypical ideas which he used in a study of the history of medicine. For a somewhat similar notion, applied to the history of recent physics, see Brush (1980).
19. Sambursky (1963), p. 203.
20. Holton (1973), Holton (1978). See also Merton (1975).
21. In this respect Holton's themata are similar to what Feuer calls iso-emotional or teleological principles. Feuer (1976).
22. Holton (1978), p. 10.
23. Sambursky (1963), p. 156.
24. Wolff (1978).

Chapter 9

1. Truesdell (1980).
2. Ibid., p. 4.
3. Collingwood (1939), p. 58.
4. An example is Wightman (1951), where Harvey is portrayed as a modern, empirically minded scientist who 'cleared away all obstructions to *just* views on the subject' (p. 345, my emphasis). Fludd does not figure in Wightman's book.
5. Buchdal (1962), p. 71.
6. Kuhn (1970*a*).
7. Bachelard (1951*a*), p. 9. Quoted from Fichant and Pécheux (1971), p. 129.
8. Ibid., p. 131.
9. Fichant and Pécheux (1971), p. 89.
10. Bachelard (1951*b*), p. 27.
11. Fichant and Pécheux (1971), p. 131.
12. Butterfield (1951).
13. Butterfield (1950), p. 54. There is a remarkable discrepancy between Butterfield's strong anti-whig morals and his historiography as practised in Butterfield (1949). The Whig flavour of the latter work illustrates, if nothing else, the difficulty of reconciling theory and practice.
14. Meyer (1905), p. 37. See also Weyer (1972).
15. According to Dijksterhuis (1961), p. 29, 'the fundamental law of Aristotelian dynamics [is] . . . the ancient analogue of the fundamental formula $F=m\cdot a$ of classical mechanics'.

16. Cohen (1977).
17. Newton (1966), p. 13.
18. Cohen (1977), p. 346.
19. Skinner (1969). While Skinner deals primarily with the historiography of political ideas, Lindholm has applied a similar criticism to the historiography of science. What Lindholm calls 'the assumption of clarity' is largely identical to Skinner's 'mythology of coherence'. Lindholm (1981).
20. Fermia (1981).
21. Westfall (1958).
22. Figala (1978), p. 108. While Figala wants to make Newton a predecessor of Bohr, the Russian scientist Vavilov saw him as a predecessor of Rutherford. 'Thus, we have sufficient grounds to believe that Newton had a good idea of the complexity of the chemical atom and even conjectured the existence of a tiny exceedingly stable atomic nucleus. In this sense Newton was a predecessor of Rutherford.' Vavilov (1947), p. 55.
23. Lindholm (1981).
24. Kuhn (1984b), p. 233 and 236.
25. Ibid. The notion of the sleepwalking discoverer was argued in Koestler (1959).
26. Skinner (1969). Canguilhem (1979). Sandler (1979).
27. Quotation taken from Canguilhem (1979), p. 20. Biot objected to the claim that the chemical revolution did not originate with Lavoisier but was anticipated by earlier researchers.
28. Glass, Temkin and Straus (1968), p. 172, here quoted from Sandler (1979), p. 189.
29. Cf. Heimann and McGuire (1971). Boscovich's main work. *Theoria Philosophiae Naturalis*, was published in 1758.
30. Agassi (1963), p. 32.
31. Duhem (1974), p. 221.
32. Merton (1975). See also Hull (1979) and Hall (1983).
33. 'Only in view of what happened later can we say that these partial statements even deal with the same aspect of nature.' Kuhn (1977), p. 70. '. . . the "simultaneous discoverers" discovered very different things, and it is only under the influence of hindsight gained from their pooled results, that their discoveries seem identical.' Elkana (1974), p. 178.
34. Butterfield (1951), p. 72.
35. Hull (1979).
36. Brannigan (1981), p. 112. Olby (1979).
37. Hooykaas (1970, p. 45.

Chapter 10

1. Althusser (1975), 52. A comprehensive discussion of the concept of ideology can be found in Plamenatz (1970).

2. Cf. Graham, Lepenies and Weingart (1983), pp. IX–XX.

3. Reprinted in Coleman (1981).

4. Lenard (1937). Lenard's Nazi history of science was parallelled in chemistry by the distinguished chemist and historian of chemistry Paul Walden, see Walden (1944).

5. Joravsky (1955). Graham (1972), chapter 8.

6. The classical examples are Draper (1875) and White (1955).

7. Jaki (1978*a*). Also Jaki (1978*b*) in which the author argues at length that 'the road of science, both historically and philosophically, is a logical access to the ways to God. The study of that road is the historiography of science'. p. 4.

8. Jaki (1978*a*), pp. 57, 78 and 61.

9. Kohlstedt and Rossiter (1985).

10. As to the transmission of science to the Third World, see Pyenson (1982). During the past few decades the development of science in the Eastern cultures has attracted many historians of science and given rise to its own journals, including *Journal of the History of Arabic Science, Historia Scientarum* (formerly *Japanese Journal of the History of Science*) and *Indian Journal for the History of Science*. Latin American science, too, has got its own journal, named *Quipu*.

11. The feeling of unfair crediting is distinct in the official *Romanian Review*, 35 (1981), special issue dedicated to the Sixteenth International Congress of the History of Science held in Bucharest in 1981. Although the neutrality and internationalism of science is constantly emphasized, the Rumanian authors also claim that a number of important discoveries were first made by Rumanian scientists and thus 'belong' to Rumania.

12. Fisher (1966), p. 158.

13. Cf. Laudan (1983).

14. Young (1802), p. 12. Young's references to Newton should not be interpreted solely as an attempt to legitimize his theory. The 'Newtonian censorship' explanation of why Young's theory did not achieve immediate recognition exaggerates the authority of Newtonianism. See Cantor (1983), pp. 129–146 and Worrall (1976), pp. 112–114.

15. According to Tait's biographer, the myth was made up as follows: '"The Conservation of Energy," he [Tait] said to Thomson one day, "must be in Newton somewhere if we can only find it." They set themselves to re-read carefully the *Principia* in the original Latin, and ere long discovered the treasure in the finishing sentences of the Scholium to Lex III.' Knott (1911), as quoted in Elkana (1974), p. 49.

16. Bensaude–Vincent (1983).

17. Porter (1976). Laudan (1983).

18. Einstein (1933), p. 1.

19. Byrne (1980). Byrne's analysis of what he calls 'Albert Einstein's theory of the history of science' exaggerates Einstein's interest and competence in

history. Einstein was a great physicist, but he was no historian. Byrne's primary source for his interpretation is Einstein and Infeld (1938) which is a popular, semi-historical work. In regard of the fact that the authors meant it to be 'a chat' (preface) it is far-fetched to analyse it as if it were a serious contribution to the history of science.

20. In a conversation with Robert Shankland in 1950. Quoted from Holton (1973), p. 327.
21. Forman (1969).
22. Ibid., p. 67.
23. Forman's provocative anthropological perspective is related to the programme which the so-called Edinburgh School developed a few years later. ' . . . those who study natural knowledge will feel free to experiment with any of the general methods and theories of the social sciences. In so far as such methods and theories appear to have merit in the context of art or religion, or the cosmologies of preliterate societies, or any other setting, they may prove useful in the study of science also. As a typical form of culture, science should be amenable to whatever methods advance our understanding of culture generally'. Barnes and Shapin (1979), p. 10.
24. Forman (1969), pp. 69–70. Forman's critical attitude towards the scientist-as-historian has not weakened over the years. Thus in an essay review of 1983: ' . . . for scientists history is not the field upon which they wrestle for truth, but principally their field of celebration and self-congratualtion'. Forman (1983), p. 826.
25. Ewald (1969), pp. 72–81.
26. Forman (1969), p. 41 and p. 68, Ewald (1969), p. 81.

Chapter 11

1. Quoted from Weyer (1974). p. 3.
2. Dahl (1967).
3. Knight (1975).
4. Amazingly, some archivists and historians think that such sources are superfluous, that 'test and experimental data should be destroyed when the information they contain is condensed in published reports or statistical summaries'. M.J.Brichford as quoted in Elliot (1974), p. 30.
5. For a recent guide to archives, bibliographies, catalogues, handbooks, etc., see Jayawardene (1982).
6. The contexts of discovery and justification are discussed in most books on theory of science. See, e.g. Lakatos and Musgrave (1970).
7. Quotation from Koestler (1960), p. 124. A modern parallel is contained in Einstein's fundamental paper on cosmology from 1917, 'Kosmologische Betrachtungen zur allgemeinen Relativitätstheorie'. In the introduction Einstein tells the reader that 'I shall conduct the reader over the road that I have myself travelled, rather a rough and winding road, because otherwise

I cannot hope that he will take much interest in the result at the end of the journey'. English translation in Einstein *et al.* (1923), p. 179.

8. Translated from an early book about electromagnetism, written by the German physicist and physiologist Paul Erman shortly after Ørsted's discovery. Quoted from Caneva (1978), p. 83.

9. Holton (1973), pp. 192–218.

10. Quoted from Frankel (1976), p. 307.

11. Cf. Hill (1975). For natural history illustrations, see Knight (1985).

12. Two examples are Harrison (1978) and Crosland (1981).

13. *Isis Critical Bibliography* (annually). Also *Bulletin Signalétique* (three times a year).

14. Merton (1957). A better known scientific fraud is the Piltdown case. The Piltdown fossils were 'discovered' in 1912 and for 40 years were generally accepted as evidence of a prehistoric man. Only in 1953 was it realized that the Piltdown man was a carefully planned fraud. See the discussion in Brannigan (1981), pp. 133–142. The Piltdown man was a forgery of the science of palaeoanthropology, not really of the history of science. In this respect it differs from the Vrain–Lucas case.

15. Bloch (1953), pp. 90–100.

16. Sarton (1936), p. 13.

17. Bloch (1953), p. 120.

Chapter 12

1. Popper (1976), p. 23.

2. Kuhn (1970b). Much of the discussion in modern theory of science deals with the degree of translatability between theories. Kuhn, Quine and Feyerabend claim that scientific theories are sometimes radically intranslatable while Popper only accepts intranslatability in a much weaker form. As to translations of texts most theorists of science, Popper included, agree with Kuhn in the following statement: 'Translation, in short, always involves compromises which alter communication. The translator must decide what alterations are acceptable. To do that he needs to know what aspects of the original it is most important to preserve and also something about the prior education and experience of those who will read his work. Not surprisingly, therefore, it is today a deep and open question what a perfect translation would be and how nearly an actual translation can approach the ideal.' Ibid., p. 268.

3. Cf. Pearce Williams (1975).

4. White (1955), p. 127. White gives no source for what appears to be Calvin's quotation.

5. Hooykaas (1973), p. 121. Rosen (1960).

6. Quoted from Ross (1962), p. 68. The source is Locke's *Essay on Human Understanding*.

7. Crosland (1978), p. 5.
8. Thackray (1966), p. 43.
9. Thomson (1830–1831). Vol.2, p. 291.
10. Thomson (1825), vol.1, p. 10. Here quoted from Thackray (1966), p. 52. My emphasis.
11. Guerlac (1961).
12. Greenaway (1958), p. 79.
13. Meldrum (1910–1911).
14. This was Partington's conclusion after having sought in vain to reproduce experimentally the data reported by Dalton. Partington (1939), p. 279.
15. Nash (1956), p. 108.
16. Thackray (1966), p. 51. See also Thackray (1972) which includes a critique of 'the continuing desire to explain Dalton's work in terms of crucial events and decisive breaks'. p. 40.
17. In addition to the titles mentioned below, see, e.g. Shea (1972) and Drake (1970).
18. Galilei (1963), p. 145.
19. Thomas Salusbury's translation from 1661, cf. Cohen (1977), p. 340.
20. Galilei (1914), p. 153.
21. Seeger (1965), p. 689.
22. See also Koyré's collection of articles in Koyré (1968).
23. Koyré (1968), p. 13. Koyré's anti-empiricist view of the scientific revolution was rooted in philosophical as well as historical considerations. He was convinced that the 'Platonic method' did not just happen to be the method of Galileo and Newton but that it constitutes the very essence of modern science: 'I do indeed believe that science is primarily theory and not gathering of "facts".' Ibid., p. 18.
24. Dijksterhuis (1961), p. 336.
25. Hall (1963), p. 34.
26. Truesdell (1968), p. 307.
27. Quoted from Drake (1978), p. 294.
28. According to Dijksterhuis (1961), p. 353, Salviati's answer is a blow against 'the myth that Galileo was a great advocate of experiment'. Dijksterhuis mentions the letter to Ingoli but apparently he does not find it trustworthy.
29. Drake and MacLachlan (1975).
30. Ibid., p. 110.
31. Hall (1963), p. 34. Also Koyré (1968), p. 94.
32. Drake (1975).
33. Ronald Naylor carefully repeated the experiments reported by Galileo but was unable to confirm Galileo's results. He then concluded that Galileo probably did not obtain his results by means of the inclining plane experiments. Naylor (1974). Shea, another Galileo scholar, summarized the ongoing discussion as follows: 'Koyré concluded that such experiments

[balls on inclined planes with sufficiently exact timekeeping] were beyond the art of experimentation, and, since Settle showed that Galileo could have performed the inclined plane experiment, Drake concluded that he actually achieved Settle's results. But surely the most we can claim for Settle's experiments, if they faithfully reproduce Galileo's, is that Galileo could have secured identical results but not that he necessarily did.' Shea (1977), p. 85. Now it is correct that the historian is not entitled to conclude that an experiment was actually performed just because it can be repeated (see also chapter 14). But neither can he conclude that moderate disagreements, such as those demonstrated by Naylor, prove that the experiment did not take place as reported. This is what Shea does.

34. Truesdell (1968), p. 307.
35. Quoted from Drake (1978), p. 19.
36. Ibid., p. 20 and p. 415. But see also the classic work of Cooper (1935) which provides a detailed examination of the sources, including Viviani's account. Cooper concluded that Galileo did not while teaching at Pisa make the alleged experiment.
37. Lodge (1960), p. 90.
38. Cf. Segre (1980).

Chapter 13

1. Holton (1969*a*), Holton (1969*b*).
2. The earliest evidence is contained in a series of conversations the psychologist Max Wertheimer had with Einstein in 1916 and later. See Wertheimer (1959), pp. 213–233. As far as the role of the Michelson experiment is concerned, Wertheimer's account is consistent with Holton's conclusions.
3. Jaffe (1960), p. 167. Jaffe's quotation is from *Science*, 73, (1931), p. 375.
4. Jaffe (1960), p. 101.
5. Quoted from Holton (1969*b*), p. 175.
6. Shankland (1963), p. 47.
7. Ibid., p. 55.
8. Quoted from Holton (1969*a*), p. 969.
9. Einstein (1982), p. 46. Translation of notes taken by J.Ishiwara who attended Einstein's speech, delivered in German at Kyoto University on 14 December 1922.
10. Quoted from Duhem (1974), p. 196.
11. Ibid.
12. Ibid., p. 198.
13. Woolgar (1976). Pulsars are celestial objects which emit rapidly pulsating radio signals of an accurate frequency. They were first detected in 1967 by a group of astronomers at Cambridge University.

14. Ibid., p. 400.
15. Ibid., p. 399.

Chapter 14

1. Belloni (1970), p. 158.
2. Greenaway (1958), p. 96.
3. Dijksterhuis (1961), p. 340. For objections to the use of Lakatosian reconstructionism, see McMullin (1970) and Holton (1978).
4. Galilei (1974), p. 74.
5. Koyré (1968), p. 84.
6. Ibid.
7. MacLachlan (1973).
8. Fogh (1921).
9. Translated after the facsimile in Kjølsen (1965), p. 105. The quotation reproduced here refers only to the first step of the aluminium synthesis, the production of non-aqueous aluminium chloride. In the second step the chloride was transformed into an amalgam by dissolving it in mercury. The metal was obtained in pure form if the amalgam was distilled.
10. Quoted frrom Kjølsen (1965), p. 108. The content of the letter to Schweigger, including its reference to carbon, was in the same year published in Poggendorf's *Annalen der Physik und Chemie*.
11. Broad (1981).
12. Elliott (1974), p. 27. See also volume 53 of *Isis* where W.C.Grover defends the opposite point of view (on p. 58).
13. Stephenson (1982).
14. Grosser (1979), p. 41 and p. 139.

Chapter 15

1. See Hankins (1979). Notable examples of modern scientific biographies which have helped to reverse the trend include Drake (1978), Manuel (1980), Westfall (1980) and Morselli (1984).
2. Agassi (1963).
3. According to E.T.Bell's widely read *Men of Mathematics*, first published in 1937. Here quoted from Rothman (1982), p. 112.
4. Rothman (1982), p. 120.
5. Manuel (1980). Westfall (1980), pp. 600–601.
6. Hankins (1979), p. 5.
7. According to Shore (1981), p. 95.
8. Feuer (1974).

Chapter 16.

1. Stone (1971) helped to make historians of science rediscover the prosopographical method.

2. Cf. Cowan (1972).
3. Ostwald (1909).
4. Cf. Mikulinsky (1974).
5. Poggendorf (1863–1976). For other biographical dictionaries, see Jayawardene (1982).
6. Ben–David and Collins (1966). Mullins (1972). Fisher (1966), Fisher (1967). Edge and Mulkay (1976). Lemaine *et al.* (1976).
7. More often than not, master–pupil networks are misleading, especially when they cover several generations. For an example, see Pledge (1959), p. 200, which seems to imply that Max Planck (1858–1947) was somehow connected with Claude–Louis Berthollet (1748–1822).
8. Fisher (1966), Fisher (1967).
9. Fisher's studies are criticized by Fang and Takayama (1975), pp. 227–238. 'The number, large or small, of those engaged in a certain theory has never been a sound or valid criterion with which the validity of the theory may be diagnosed. . . . After all, never has mathematics been an undisciplined champion of democracy for counting noses.' p. 237. But if Fisher's data are reliable it would indeed be strange to claim that the invariant theory was a vital discipline in the period 1935–1941. Mathematics may be more aristocratic than democratic, but even an aristocracy is nothing if there are not aristocrats.
10. Pyenson (1977), p. 172.
11. Ibid., p. 179.
12. Thackray (1974). For objections to Thackray's approach, see the same journal, *80* (1975), pp 203–204.
13. Thackray (1974), p. 681 and 698.
14. Shapin and Thackray (1974).
15. Ibid., p. 21.
16. Shapin (1974), Shapin (1975).

Chapter 17

1. Cole and Eames (1917).
2. Rainoff (1929).
3. Merton (1938). Merton had earlier used quantitative methods in the history of science, see Merton and Sorokin (1935). Surveys of the development of quantitative history of science include Merton (1977) and Thackray (1978).
4. Gilbert and Woolgar (1974). Gilbert (1978). Edge (1979).
5. Cf. Narin (1978).
6. Shapin and Thackray (1974), p. 7.
7. Solla Price (1963), Solla Price (1974), Solla Price (1980).
8. Solla Price (1972).
9. Solla Price (1956), Solla Price (1974).
10. Solla Price (1974), p. 172.
11. Cf. Menard (1971).

12. Solla Price (1963), p. 41.
13. The French mathematician A.Cauchy (1789–1857) wrote almost 800 scientific publications, his Irish colleague A.Cayley (1821–1895) almost 1,000, and the French chemist M.Berthelot (1827–1907) was the author or co-author of no less than 1,600 publications. However, these examples of highly productive and eminent scientists are exceptional. Theodore Cockerell (1866–1948), a professor in natural history, published 3,904 works during his lifetime. Although not recorded in *Guinness Book of Records*, this is probably a world record in publication mania.
14. The diagram appears in Solla Price (1963), p. 29, Rescher (1978), p. 169 and Dobrov (1969), p. 66.
15. Dobrov (1969), p. 66.
16. See, e.g. Yuasa (1962) who concludes that the dominant centres of science since the 16th century have moved from Italy, via England, France and Germany to the USA. Hardly a discovery that needs the kind of quantitative support provided by Yuasa.
17. Simonton (1976). Notice the difference in perspective adopted by Solla Price and Simonton. While Solla Price wants to determine the overall influence of wars on scientific production, Simonton's analysis is made 'with the aim of determining the specific causal relation between war on the societal level and scientific discovery on the individual level.' Ibid., p. 135. For an elaborated version of Simonton's quantitative science studies, see Simonton (1984).
18. Taken from Sorokin (1937). Sorokin's table is mainly based on data from Darmstaedter (1906).
19. Solla Price (1980).
20. Ibid., p. 180.
21. Ibid.
22. In addition Solla Price draws the following, more far-reaching conclusions from the data. (1) The Industrial Revolution is not an objective historical reality but rather a convenient label due to the arbitrary periodization used by the historians. (2) Although a reality, the Scientific Revolution does not mark the beginning of the development of modern science; Copernicus, Galileo, Kepler and Boyle were predecessors of the scientific take-off which only took place at the end of the 18th century. (3) Contrary to what is often stated, the take-off phases of chemistry and biology were not delayed; it was astronomy and mechanics which began their development at an exceptionally early date.
23. Cf. Brannigan (1981).
24. CSI consists of several sections. In addition to the *Citation Index* there is a *Source Index*, including new publications, and a *Permuterm Subject Index* which classify the contributions according to specialty and codewords.
25. A first example is Small (1981) which counts 21,000 publications and

167,000 references. However, the volumes are of limited value to the historian of physics. The editor has made a very narrow selection of periodicals, excluding journals such as *Comptes rendus, Nature* and *Die Naturwissenschaften*. The selection is presumably made because these journals were not exclusively devoted to physics. Still it is a fact that many important contributions to physics appeared in these journals. The exclusion of most journals associated with applied physics or interdisciplinary physics adds to the general bias of the work.

26. Moravcsik and Murugesan (1975).
27. Quoted from Forman (1973), p. 157.
28. Merton (1957).
29. Gaston (1971) p. 486.
30. Concerning Mendel's discovery and its fate, see Zirkle (1964), Olby (1966) and Vorzimmer (1968).
31. Darwin's reading habits included his adding a mark of his own to indicate that he had looked through a work. This mark does not appear on Focke's book.
32. Vorzimmer (1968), p. 81.
33. Sullivan, White and Barboni (1977*a*), Sullivan, White and Barboni (1977*b*), Sullivan, White and Barboni (1979).
34. Lakatos and Musgrave (1970), pp. 91–196.
35. Sullivan, White and Barboni (1979), p. 323. The study of Edge and Mulkay mentioned in the quotation is Edge and Mulkay (1976).
36. Sullivan, White and Barboni (1977*b*). Small (1977).
37. Edge (1979), p. 115.

Bibliography

Agassi, J. (1963). *Towards an Historiography of Science*. s'Gravenhage: Mouton & Co. (Beiheft 2 of *History and Theory*)

Agassi, J. and Cohen, R.S., eds. (1981). *Scientific Philosophy Today*. Dordrecht: D.Reidel.

Althusser, L. (1975). *Filosofi, Ideologi og Videnskab*. Copenhagen: Rhodos (Danish translation of *Philosophie et philosophie spontanée des savants*, Paris: Maspero, 1974)

Andreski, S., ed. (1974). *The Essential Comte*. London: Croom Helm.

Atkinson, R.F. (1978). *Knowledge and Explanation in History*. London: MacMillan.

Bachelard, G. (1951a). *L'actualité de l'histoire des sciences*. Paris: Palais de la découverte.

Bachelard, G. (1951b). *L'activité rationaliste de la physique contemporaine*. Paris: Presses Universitaires de France.

Badash, L. (1965). 'Chance favors the prepared mind', *Archives Internationale d'Histoires des Sciences*, 18, 55–66.

Bailly, J.S. (1782). *Histoire de l'astronomie moderne*. 3 vols., Paris.

Barnes, B., ed. (1972). *Sociology of Science*. Harmondsworth: Penguin.

Barnes, B. and Shapin, S., eds. (1979). *Natural Order*. London: Sage Publications.

Beard, C.A. (1935). 'That noble dream', *The American Historical Review*, 41, 74–87. Reprinted in Stern (1956), pp. 315–328.

Belloni, L. (1970). 'The repetition of experiments and observations: its value in studying the history of medicine (and science)', *Journal of the History of Medicine and Allied Sciences*, 25, 158–167.

Ben–David, J, and Collins, R. (1966). 'Social factors in the origin of a new science: the case of psychology,' *American Sociological Review*, 31, 451–465.

Bensaude–Vincent, B. (1983). 'A founder myth in the history of sciences? The Lavoisier case.' In Graham, Lepenies and Weingart (1983), pp. 53–78.

Bernal, J.D. (1969). *Science in History*. 4 vols., Harmondsworth: Pelican.

Beyerchen, A.D. (1977). *Scientists Under Hitler: Politics and the Physics Community in the Third Reich*. New Haven: Yale University Press.

Blake, C. (1959). 'Can history be objective?' In Gardiner (1959), pp. 329–343.

Bloch, M. (1953). *The Historian's Craft*. New York: Vintage Books.

Bloor, D. (1976). *Knowledge and Social Imagery*. London: Routledge and Kegan Paul.

Blüh, O. (1968). 'Ernst Mach as an historian of physics,' *Centaurus, 13*, 62–84.

Boas, M. and Hall, A. Rupert (1958). 'Newton's chemical experiments,' *Archives Internationale d'Histoires des Sciences, 11*, 113–152.

Bonelli, M. and Shea, W.R., eds. (1975). *Reason, Experiment, and Mysticism in the Scientific Revolution*. New York: Science History Publications.

Brannigan, A. (1981). *The Social Basis of Scientific Discoveries*. Cambridge: Cambridge University Press.

Brewster, D. (1855). *Memoirs of the Life, Writings, and Discoveries of Sir Isaac Newton*. 2 vols., Edinburgh.

Broad, W.J. (1981). 'Sir Isaac Newton: mad as a hatter.' *Science, 213*, 1341–1344.

Brush, S.G. and King, A.L., eds. (1972). *History in the Teaching of Physics*. Hanover (New Hampshire): University Press of New England.

Brush, S.G. (1974). 'Should the history of science be rated X?' *Science, 183*, 1164–1172.

Brush, S.G. (1980). 'The Chimerical Cat: philosophy of quantum mechanics in historical perspective.' *Social Studies of Science, 10*, 393–447.

Buchdal, G. (1962). 'On the presuppositions of historians of science.' *History of Science, 1*, 67–77.

Butterfield, H. (1949). *The Origins of Modern Science, 1300–1800*. London: G.Bell & Sons.

Butterfield, H. (1950). 'The historian and the history of science.' *Bulletin of the British Society for the History of Science, 1*, 49–57.

Butterfield, H. (1951). *The Whig Interpretation of history*. New York: Charles Scribner's Sons. First published London, 1931.

Butts, R.E. and Hintikka, J., eds. (1977). *Historical and Philosophical Dimensions of Logic, Methodology and Philosophy of Science*. Dordrecht: D.Reidel.

Byrne, P.H. (1980). 'The significance of Einstein's use of the history of science.' *Dialectica, 34*, 263–274.

Caneva, K.L. (1978). 'From galvanism to electrodynamics: the transformation of German physics and its social context.' *Historical Studies in the Physical Sciences, 9*, 63–159.

Canguilhem, C. (1979). *Wissenschaftsgeschichte und Epistemologie*. Frankfurt: Suhrkamp.

Cantor, G.N. (1983). *Optics after Newton*. Manchester: Manchester University Press.

Carr, E.H. (1968). *What is History?* Harmondsworth: Penguin.

Childe, G. (1964). *What Happened in History*. Harmondsworth: Penguin.

Clark, J.T. (1971). 'The science of history and the history of science.' In Roller (1971), pp. 283–296.

Cohen, I.B. (1963). 'History of science as an academic discipline.' In Crombie (1961), pp. 769–780.

Cohen, I.B. (1976). 'The eighteenth-century origins of the concept of scientific revolution'. *Journal of the History of Ideas, 37*, 257–288.

Cohen, I.B. (1977). 'History and philosophy of science.' In Suppe (1977), pp. 308–349.

Cole, F.J. and Eames, N.B. (1917). 'The history of comparative anatomy: a statistical analysis of the literature.' *Science Progress, 11*, 578–596.

Coleman, W., ed. (1981). *French Views on German Science*. New York: Arno.

Collingwood, R.G. (1939). *An Autobiography*. Oxford: Oxford University Press.

Collingwood, R.G. (1980). *The Idea of History*. Oxford: Oxford University Press.

Collins, H.M. and Cox, G. (1976). 'Recovering relativity: did prophecy fail?' *Social Studies of Science, 6*, 423–444.

Conant, J.B. (1961). *Science and Common Sense*. Clinton (Mass.): Yale University Press.

Cooper, L. (1935). *Aristotle, Galileo, and the Tower of Pisa*. Port Washington, New York: Kennikat Press.

Corsi, P. and Weindling, P., eds. (1983). *Information Sources in the History of Science and Medicine*. London: Butterworths.

Cowan, R.S. (1972). 'Francis Galton's statistical ideas: the influence of eugenics.' *Isis, 63*, 509–528.

Crane, D. (1972). *Invisible Colleges*. Chicago: Chicago University Press.

Croce, B. (1941). *History as the Story of Liberty*. London: Allen and Unwin.

Crombie, A.C. (1953). *Robert Grosseteste and the Origins of Experimental Science, 1100–1700*. Oxford: Clarendon Press.

Crombie, A.C., ed. (1961). *Scientific Change*. London: Heinemann.

Crosland, M.P. (1978). *Historical Studies in the Language of Chemistry*. New York: Dover.

Crosland, M.P. (1981). 'The library of Gay-Lussac.' *Ambix, 28*, 158–170.

Dahl, O. (1967). *Grunntrekk i Historieforskningens Metodelære*. Oslo: Universitetsforlaget.

Daniel, G. (1980). 'Megalithic Monuments.' *Scientific American*, July, 64–76.

Dannemann, F. (1906). *Quellenbuch zur Geschichte der Naturwissenschaft in Deutschland*. Leipzig.

Dannemann, F. (1910–1913). *Die Naturwissenschaften in ihrer Entwicklung und in ihrem Zusammenhang*. 4 vols., Leipzig.

Danto, A.C. (1965). *Analytical Philosophy of History*. Cambridge: Cambridge University Press.

Darmstaedter, L. (1906). *Handbuch zur Geschichte der Naturwissenschaften und der Technik*. Berlin.

Darwin, C. (1872). *On the Origin of Species*. 6th edn. London.

Dewey, J. (1949). *Logic, the Theory of Inquiry*. New York: H. Holt and Co.

Diederich, W., ed. (1974). *Theorien der Wissenschaftsgeschichte*. Frankfurt: Suhrkamp.

Dijksterhuis, E.J. (1961). *The Mechanization of the World Picture*. London: Oxford University Press.

Dobbs, B.J.T. (1975). *The Foundation of Newton's Alchemy. Or 'the Hunting of the Greene Lyon'*. Cambridge: Cambridge University Press.

Dobrov, G.M. (1969). *Wissenschaftswissenschaft*. Berlin (DDR): Akademie–Verlag.

Dolby, R.G.A. (1977). 'The transmission of science.' *History of Science, 15*, 1–43.

Dolby, R.G.A. (1980). 'Controversy and consensus in the growth of scientific knowledge.' *Nature and System, 2*, 199–218.

Drake, S. (1970). *Galileo Studies: Personality, Tradition, and Revolution*. Ann Arbor (Mass,): University of Michigan Press.

Drake, S. (1975). 'The role of music in Galileo's experiments.' *Scientific American*, June, 98–104.

Drake, S. (1978). *Galileo at Work*. Chicago: Chicago University Press.

Drake, S. and MacLachlan, J. (1975). 'Galileo's discovery of the parabolic trajectory.' *Scientific American*, March, 102–110.

Draper, J.W. (1875). *History of the Conflict between Religion and Science*. New York.

Dray, W.H. (1957). *Laws and Explanations in History*. Oxford: Oxford University Press.

Dray, W.H. (1980). *Perspectives on History*. London: Routledge and Kegan Paul.

Du Bois–Reymond, E. (1886). *Reden*. Leipzig.

Duhem, P. (1905–1907). *Les origines de la statique*. 2 vols., Paris.

Duhem, P. (1906–1913). *Études sur Léonard de Vinci*. 3 vols., Paris.

Duhem, P. (1913–1959). *Le système du monde*. 10 vols., Paris.

Duhem, P. (1974). *The Aim and Structure of Physical Theory*. New York: Atheneum. First published Paris, 1906.

Durbin, P.T., ed. (1980). *A Guide to the Culture of Science, Technology and Medicine*. New York: Free Press.

Edge, D. (1979). 'Quantitative measures of communication in science: a critical review.' *History of Science, 17*, 102–134.

Edge, D. and Mulkay, M.J. (1976). *Astronomy Transformed. The Emergence of Radioastronomy in Britain*. New York: John Wiley.

Einstein, A. *et al.* (1923). *The Principle of Relativity*. London: Methuen.

Einstein, A. (1933). *On the Method of Theoretical Physics*. Oxford: Clarendon Press.

Einstein, A. (1982). 'How I created the theory of relativity.' *Physics Today*, August, 45–47.

Einstein, A. and Infeld, L. (1938). *The Evolution of Physics*. New York: Simon and Schuster.

Elkana, Y. (1974). *The Discovery of the Conservation of Energy*. London: Hutchison.

Elkana, Y. (1977). 'The historical roots of modern physics.' In Weiner (1977), pp. 197–265.

Elkana, Y. *et al.*, eds. (1978). *Towards a Metric of Science*. New York: John Wiley.

Elliott, C.A. (1974). 'Experimental data as a source for the history of science.' *American Archivists, 37,* 27–35.

Elzinga, A. (1979). 'The growth of knowledge.' University of Gothenburg, Department of Theory of Science, Report no. 16.

Engelhardt, D. (1979). *Historisches Bewusstsein in der Naturwissenschaft.* Freiburg: Alber.

Engels, F. (1886). *Ludwig Feuerbach und der Ausgang der klassischen deutschen Philosophie*. Stuttgart.

Ewald, P.P. (1969). 'The myth of myths: comments on P. Forman's paper.' *Archive for History of Exact Sciences, 6,* 72–81.

Fang, J. and Takayama, K.P. (1975). *Sociology of Mathematics and Mathematicians*. New York: Paideia.

Feigl, H. and Brodbeck, M., eds. (1953). *Readings in the Philosophy of Science*. New York: Appleton-Century-Crofts.

Fermia, J.V. (1981). 'An historicist critique of "revisionist" methods for studying the history of ideas.' *History and Theory, 20,* 113–134.

Feuer, L. (1974). *Einstein and the Generations of Science*. New York: Basic Books.

Feuer, L. (1976). 'Teleological principles in science.' *Inquiry, 21,* 377–407.

Fichant, M. and Pécheux, M. (1971). *Om Vetenskapernas Historia*. Stockholm: Bo Cavefors (Swedish translation of *Sur l'histoire des sciences*, Paris: Maspero, 1969).

Figala, K. (1977). 'Newton as alchemist.' *History of Science, 15,* 102–137.

Figala, K. (1978). 'Newtons rationale System der Alchemie.' *Chemie in unserer Zeit, 12,* 101–110.

Finnochiaro, M.A. (1973). *History of Science as Explanation*. Detroit: Wayne State University Press.

Fisher, C.S. (1966). 'The death of a mathematical theory: a study in the sociology of knowledge.' *Archive for History of Exact Sciences, 3,* 137–159.

Fisher, C.S. (1967). 'The last invariant theorists. A sociological study of the collective biographies of mathematical specialists.' *European Journal of Sociology, 8,* 216–244.

Fisher, N. (1982). 'Avogadro, the chemists, and historians of chemistry.' *History of Science, 20,* 77–102, 212–231.

Fleck, L. (1980). *Entstehung und Entwicklung einer wissenschaftlichen Tatsache.* Frankfurt: Suhrkamp. First published Basle, 1935.

Fogh, I. (1921). 'Über die Entdeckung des Aluminiums durch Oersted im Jahre 1825.' *Kongelige Danske Videnskabernes Selskab, Matematisk–Fysiske Meddelelser*, III, *14,* 1–17 and *15,* 1–7.

Forbes, R.J. and Dijksterhuis, E.J. (1963). *A History of Science and Technology*. 2 vols., Harmondsworth: Penguin.

Forman, P. (1969). 'The discovery of X-rays by crystals: a critique of the myths.' *Archive for History of Exact Sciences, 6,* 38–71.

Forman, P. (1973). 'Scientific internationalism and the Weimar physicists.' *Isis*, *64*, 151–178.

Forman, P. (1983). 'A venture in writing history.' *Science, 220*, 824–827.

Frankel, H. (1976). 'Alfred Wegener and the specialists.' *Centaurus, 20*, 305–324.

Frängsmyhr, T. (1973–1974). 'Science or history: Georges Sarton and the positivist tradition in the history of science.' *Lychnos*, 104–144.

Galilei, G. (1914). *Dialogues Concerning Two New Sciences*. Trans. H. Crew and A. de Salvio, New York: Macmillan. First published 1638.

Galilei, G. (1963). *Dialogues Concerning the Two Chief World Systems*. Trans. S.Drake, Berkeley: University of California Press. First published 1632.

Galilei, G. (1974). *Two New Sciences, including Centres of Gravity and Force of Percussion*. Trans. S.Drake, Madison: University of Wisconsin.

Gallie, W.B. (1964). *Philosophy and the Historical Understanding*. London: Chatto and Windus.

Gardiner, P., ed. (1959). *Theories of History*. New York: Free Press.

Gardiner, P. (1952). *The Nature of Historical Explanation*. Oxford: Oxford University Press.

Gaston, J. (1971). 'Secretiveness and competition for priority in physics.' *Minerva, 9*, 472–492.

Giere, R. (1973). 'History and philosophy of science: intimate relationship or marriage of convenience?' *British Journal for the Philosophy of Science, 24*, 282–297.

Gilbert, G.N. (1978). 'Measuring the growth of science: a review of indicators of scientific growth.' *Scientometrics, 1*, 9–34.

Gilbert, G.N. and Woolgar, S. (1974). 'The quantitative study of science: an examination of the literature.' *Science Studies, 4*, 279–294.

Gilbert, G.N. and Mulkay, M. (1984). 'Experiments are the key.' *Isis, 75*, 105–125.

Gillispie, C.C., ed. (1970–1980). *Dictionary of Scientific Biography*. 16 vols., New York: Charles Scribner's Sons.

Glass, B., Temkin, O. and Straus, W.L., eds. (1968). *Forerunners of Darwin: 1745–1859*. Baltimore: John Hopkins Press.

Goodman, N. (1955). *Fact, Fiction, and Forecast*. Cambridge, Mass.: Harvard University Press.

Gould, J.D. (1969). 'Hypothetical History.' *The Economic History Review, 22*, 195–207.

Graham, L.R. (1972). *Science and Philosophy in the Soviet Union*. New York: Alfred A. Knopf.

Graham, L., Lepenies, W. and Weingart, P., eds. (1983). *Functions and Uses of Disciplinary Histories*, Dordrecht: D.Reidel.

Greenaway, F. (1958). *The Biographical Approach to John Dalton*. Memoirs and Proceedings of the Manchester Literary and Philosophical Society, vol. 100.

Greene, M.T. (1982). *Geology in the Nineteenth Century*. Ithaca: Cornell University Press.

Greene, M.T. (1985). 'History of Geology.' *Osiris* (2), 1, 97–116.

Grimsehl, E. (1911). *Didaktik und Methodik der Physik.* Munich.

Grmek, M.D., Cohen, R.S. and Cimino, G., eds. (1980). *On Scientific Discovery.* Dordrecht: D.Reidel.

Grosser, M. (1979). *The Discovery of Neptune.* New York: Dover.

Guerlac, H. (1961). 'Some Daltonian doubts.' *Isis, 52*, 544–554.

Guerlac, H. (1963). 'Some historical assumptions of the history of science.' In Crombie (1963), pp. 797–812.

Gunter, P. (1971). 'Bergson's theory of matter and modern cosmology'. *Journal of the History of Ideas, 32*, 525–542.

Hahn, R. (1975). 'New directions in the social history of science.' *Physis, 17*, 205–218.

Hall, A. Rupert (1963). *From Galileo to Newton, 1630–1720.* London: Collins.

Hall, A. Rupert (1969). 'Can the history of science be history?' *British Journal for the History of Science, 4*, 207–220.

Hall, A. Rupert (1983). 'On Whiggism.' *History of Science, 21*, 45–59.

Hankins, T.L. (1979). 'In defence of biography: the use of biography in the history of science.' *History of Science, 17*, 1–16.

Harrison, J. (1978). *The Library of Isaac Newton.* Cambridge: Cambridge University Press.

Hayek, F.A. (1952). *The Counter-Revolution of Science.* Glencoe (Illinois): Free Press.

Heiberg, J.L. (1912). *Naturwissenschaften und Mathematik im klassischen Altertum.* Leipzig: Teubner.

Heimann, P.M. and McGuire, J.E. (1971). 'Newtonian forces and Lockean powers: concepts of matter in eighteenth-century thought.' *Historical Studies in the Physical Sciences, 3*, 233–306.

Hempel, C.G. (1942). 'The function of general laws in history.' *Journal of Philosophy, 39*, 35–48.

Hempel, C. (1965). *Aspects of Scientific Explanation.* New York: Free Press.

Hendrick, R.E. and Murphy, A. (1981). 'Atomism and the illusion of crisis: the danger of applying Kuhnian categories to current particle physics.' *Philosophy of Science, 48*, 454–468.

Hermerén, G. (1977). 'Criteria of objectivity in history.' *Danish Yearbook of Philosophy, 14*, 13–35.

Hesse, M.B. (1960). 'Gilbert and the historians.' *British Journal for the Philosophy of Science, 11*, 1–10, 131–142.

Hiebert, E.N. (1970). 'Mach's philosophical use of the history of science.' In Stuewer (1970), pp. 184–203.

Hill, C.R. (1975). 'The iconography of the laboratory.' *Ambix, 22*, 102–110.

Hoefer, F. (1842–1843). *Histoire de la chimie.* 2 vols., Paris.

Holton, G. (1969a). 'Einstein and the "crucial" experiments.' *American Journal of Physics, 37*, 968–982.

Holton, G. (1969*b*). 'Einstein, Michelson and the "crucial" experiments.' *Isis*, 60, 133–197.

Holton, G. (1973). *Thematic Origins of Scientific Thought*. Cambridge (Mass.): Harvard University Press.

Holton, G. (1978). *The Scientific Imagination: Case Studies*. Cambridge (Mass.): Harvard University Press.

Hooykaas, R. (1970). 'Historiography of science, its aim and methods.' *Organon*, 7, 37–49.

Hooykaas, R. (1973). *Religion and the Rise of Modern Science*. Edinburgh: Scottish Academic Press.

Howson, C., ed. (1976). *Method and Appraisal in the Physical Sciences*. Cambridge: Cambridge University Press.

Hull, D.L. (1979). 'In defence of presentism.' *History and Theory*, 18, 1–15.

Hunter, M. (1981). *Science and Society in Restoration England*. Cambridge: Cambridge University Press.

Jacob, J.R. (1977). *Robert Boyle and the English Revolution*. New York: Burt Franklin.

Jaffe, B. (1960). *Michelson and the Speed of Light*. New York: Doubleday & Co.

Jagnaux, R. (1891). *Histoire de la chimie*. Paris.

Jaki, S.L. (1966). *The Relevance of Physics*. Chicago: Chicago University Press.

Jaki, S.L. (1978*a*). *The Origin of Science and the Science of its Origin*. Edinburgh: Scottish Academic Press.

Jaki, S.L. (1978*b*). *The Road of Science and the Ways to God*. Edinburgh: Scottish Academic Press.

Jammer, M. (1961). *Concepts of Mass*. Cambridge (Mass.): Harvard University Press.

Jayawardene, S.A. (1982). *Reference Books for the Historian of Science*. London: Science Museum.

Joravsky, D. (1955). 'Soviet views in the history of science.' *Isis*, 46, 3–13.

Kjølsen, H.H. (1965). *Fra Skidenstræde til H.C.Ørsted Institutet*. Copenhagen: Gjellerup.

Knight, D. (1975). *Sources for the History of Science*. New York: Cornell University Press.

Knight, D. (1985). 'Scientific theory and visual language.' *Acta Universitatis Upsaliensis*, New Series, 22, 106–124.

Knott, C.G. (1911). *The Scientific Work of P.G.Tait*. Cambridge.

Koestler, A. (1959). *The Sleepwalkers*. New York: Hutchison.

Koestler, A. (1960). *The Watershed. A Biography of Johannes Kepler*. New York: Doubleday Anchor.

Kohlstedt, S.G. and Rossiter, M.W., eds. (1985). *Historical Writing on American Science*. Philadelphia: History of Science Society (volume 1 of *Osiris*, 2nd series).

Kopp, H. (1843–1847). *Geschichte der Chemie*. 4 vols., Braunschweig.

Koyré, A. (1968). *Metaphysics and Measurement*. London: Chapman and Hall.

Kracauer, S. (1966). 'Time and history.' *History and Theory*, Beiheft 6, 65–78.

Krafft, F. (1976). 'Die Naturwissenschaften und ihre Geschichte.' *Sudhoffs Archiv,* 60, 317–337.

Kröber, G. (1978). 'Wissenschaftswissenschaft und Wissenschaftsgeschichte.' *Zeitschrift für Geschichte der Naturwissenschaft, Technik und Medizin, 15,* 63–89.

Kuhn, T.S. (1970*a*). *The Structure of Scientific Revolutions.* Chicago: University of Chicago Press.

Kuhn, T.S. (1970*b*). 'Reflections on my critics.' In Lakatos and Musgrave (1970), pp. 231–278.

Kuhn, T.S. (1977). *The Essential Tension: Selected Studies in Scientific Tradition and Change.* Chicago: University of Chicago Press.

Kuhn, T.S. (1978). *Black-Body Theory and the Quantum Discontinuity.* Oxford: Clarendon Press.

Kuhn, T.S. (1984*a*). 'Professionalization recollected in tranquility.' *Isis, 75,* 29–32.

Kuhn, T.S. (1984*b*). 'Revisiting Planck.' *Historical Studies in the Physical Sciences, 14,* 231–252.

Lakatos, I. and Musgrave, A., eds. (1970). *Criticism and the Growth of Knowledge.* Cambridge: Cambridge University Press.

Lakatos, I. (1974). 'Die Geschichte der Wissenschaft und ihre rationalen Rekonstruktionen.' In Diederich (1974), pp. 55–119.

Laudan, L. (1977). *Progress and its Problems.* London: Routledge and Kegan Paul.

Laudan, R. (1983). 'Redefinitions of a discipline: histories of geology and geological history.' In Graham, Lepenies and Weingart (1983), pp. 79–104.

Leibniz, G.W. (1849–1863). *Mathematische Schriften.* Berlin.

Lemaine, G. *et al.,* eds. (1976). *Perspectives on the Emergence of Scientific Disciplines.* The Hague: Mouton & Co.

Lenard, P. (1937). *Grosse Naturforscher.* Munich: J.F. Lehman.

Lepenies, W. (1977). 'Problems of a historical study of science.' In Mendelsohn, Weingart and Whitley (1977), pp. 55–67.

Libby, W. (1914). 'The history of science.' *Science*, 40, 670–673.

Liebig, J.V. (1874). *Reden und Abhandlungen.* Leipzig.

Lilley, S. (1953). 'Cause and effect in the history of science.' *Centaurus, 3,* 58–72.

Lindholm, L.M. (1981). 'Is realistic history of science possible?' In Agassi and Cohen (1981), pp. 159–186.

Lodge, O. (1960). *Pioneers of Science.* New York: Dover. First published London, 1896.

Losee, J. (1983). 'Whewell and Mill on the relation between philosophy of science and history of science.' *Studies in History and Philosophy of Science, 14,* 113–126.

Lovejoy, A.O. (1976). *The Great Chain of Being.* Cambridge (Mass.): Harvard University Press. First published Cambridge (Mass.), 1936.

Mach, E. (1960). *The Science of Mechanics: A Critical and Historical Account of its Development*. LaSalle (Illinois): Open Court. First published Leipzig, 1883.

MacLachlan, J. (1973). 'A test of an "imaginary" experiment of Galileo's.' *Isis*, 64, 374–379.

Mandelbaum, M. (1971). *The Problem of Historical Knowledge*. New York: Books for Libraries Press.

Mann, G. (1980). 'Geschichte als Wissenschaft und Wissenschaftsgeschichte bei Du Bois–Reymond.' *Historische Zeitschrift, 231*, 75–100.

Manuel, F.E. (1980). *A Portrait of Isaac Newton*. London: Frederick Muller.

Marwick, A. (1970). *The Nature of History*. London: MacMillan.

Marx, K. and Engels, F. (1971). *Karl Marx og Friedrich Engels. Udvalgte Skrifter*. 2 vols., Copenhagen: Tidens Forlag.

McMullin, E. (1970). 'The history and philosophy of science: a taxonomy.' In Stuewer (1970), pp. 12–67.

Meldrum, A.N. (1910–1911). *The Development of the Atomic Theory*. Memoirs and Proceedings of the Manchester Literary and Philosophical Society, vols, 54 and 55.

Menard, H.W. (1971). *Science: Growth and Change*. Cambridge (Mass.): Harvard University Press.

Mendelsohn, E., Weingart, P. and Whitley, R., eds. (1977): *The Social Production of Scientific Knowledge*. Dordrecht: D.Reidel.

Merton, R.K. (1938). 'Science, technology, and society in seventeenth-century England.' *Osiris, 4*, 360–632. Reprinted, New York: Humanities Press, 1978.

Merton, R.K. (1957). 'Priorities in scientific discovery: a chapter in the sociology of science.' *American Sociological Review, 22*, 635–659.

Merton, R.K. (1975). 'Thematic analysis in science.' *Science, 188*, 335–338.

Merton, R.K. (1977). 'The sociology of science: an episodic memoir.' In Merton and Gaston (1977), pp. 3–141.

Merton, R.K. and Sorokin, P.A. (1935). 'The course of Arabian intellectual development, 700–1300 A.D.' *Isis, 22*, 516–524.

Merton, R.K. and Gaston, J., eds. (1977). *The Sociology of Science in Europe*. Carbondale (Illinois): Southern Illinois University Press.

Merz, J.T. (1896–1914). *A History of European Thought in the Nineteenth Century*. 4 vols., London. Reprinted, New York: Dover, 1965.

Meyer, E.V. (1905). *Geschichte der Chemie*. Leipzig.

Mikulinsky, S. (1974). 'Alphonse de Candolle's *Histoire des sciences et des savants depuis deux siècles* and its historical significance.' *Organon, 10*, 223–243.

Mikulinsky, S. (1975). 'The methodological problems of the history of science.' *Scientia, 110*, 83–97.

Mill, J.S. (1843). *A System of Logic*. 2 vols., London.

Moravcsik, M.J. and Murugesan, P. (1975). 'Some results on the function and quality of citations.' *Social Studies of Science, 5*, 86–92.

Morselli, M. (1984). *Amedeo Avogadro. A Scientific Biography*. Dordrecht: Reidel.

Mullins, N.C. (1972). 'The development of a scientific specialty: the phage group and the origins of molecular biology.' *Minerva, 10*, 51–82.

Nagel, E. (1961). *The Structure of Science*. London: Routledge and Kegan Paul.

Narin, F. (1978). 'Objectivity versus relevance in studies of scientific advance.' *Scientometrics, 1*, 35–41.

Nash, L.K. (1956). 'The origin of Dalton's chemical atomic theory.' *Isis, 47*, 101–116.

Naylor, R. (1974). 'Galileo and the problem of the free fall.' *British Journal for the History of Science, 7*, 10–134.

Needham, J. (1943). *Time: The Refreshing River*. London: Allen and Unwin.

Newton, I. (1966). *Principia*. Berkeley: University of California Press (Motte's translation, first published London, 1729).

Nietzsche, F. (1874). *Unzeitgemässe Betrachtungen: Vom Nutzen und Nachteil der Historie für das Leben.* Leipzig.

Nye, M.J. (1981). 'N-rays: an episode in the history and psychology of science.' *Historical Studies in the Physical Sciences, 11*, 125–156.

Oakeshott, M.J. (1933). *Experience and its Modes*. Cambridge: Cambridge University Press.

Olby, R.C. (1966). *The Origins of Mendelism*. London: Constable.

Olby, R.C. (1979). 'Mendel no Mendelian?' *History of Science, 17*, 53–72.

Olszewski, E. (1964). 'Periodization of the history of science and technology.' *Organon, 1*, 195–206.

Ostwald, W., ed.(1889). *Klassiker der Exakten Naturwissenschaften*. Leipzig.

Ostwald, W. (1909). *Grosse Männer*. Leipzig.

Ørsted, H.C. (1856). *Ånden i Naturen*. Copenhagen. First published Copenhagen, 1851 (English translation: *The Soul in Nature*, London, 1852).

Pais, A. (1982). *Subtle is the Lord. The Science and the Life of Albert Einstein*. Oxford: Oxford University Press.

Partingon, J.R. (1939). 'The origin of the atomic theory.' *Annals of Science, 4*, 245–282.

Paul, H.W. (1976). 'Scholarship versus ideology: the chair of the general history of science at the Collège de France, 1892–1913.' *Isis, 67*, 376–398.

Pearce Williams, L. (1966a). 'The historiography of Victorian science.' *Victorian Studies, 9*, 197–204.

Pearce Williams, L. (1966b). (Letter to the Editor) *Scientific American*, June.

Pearce Williams, L. (1975). 'Should philosophers be allowed to write history?' *British Journal for the Philosophy of Science, 26*, 241–253.

Pedersen, O. (1975). *Matematik og Naturbeskrivelse i Oldtiden*. Copenhagen: Akademisk Forlag.

Pelling, M. (1983). 'Medicine since 1500.' In Corsi and Weindling (1983), pp. 379–409.

Plamenatz, J. (1970). *Ideology*. London: Pall Mall Press.

Pledge, H.T. (1959). *Science since 1500*. New York: Harpers.

Poggendorf, J.C. (1863–1976). *Biographisch–literarisches Handwörterbuch zur Geschichte der exakten Wissenschaften*. 7 vols., Leipzig: Akademie Verlag.

Popper, K.R. (1961). *The Poverty of Historicism*. London: Routledge and Kegan Paul.

Popper, K.R. (1969). 'A pluralist approach to the philosophy of science.' In Streisler *et al.* (1969), pp. 181–200.

Popper, K.R. (1976). *Unended Quest. An Intellectual Autobiography*. Glasgow: Fontana.

Porter, R. (1976). 'Charles Lyell and the principles of the history of geology.' *British Journal for the History of Science, 9*, 91–103.

Priestley, J. (1775). *The History and Present State of Electricity*. London. First published London, 1767.

Pyenson, L. (1977). '"Who the guys were": prosopography in the history of science.' *History of Science, 15*, 155–188.

Pyenson, L. (1982). 'Cultural imperialism and exact sciences.' *History of Science, 20*, 1–43.

Rainoff, T.J. (1929). 'Wave-like fluctuations of creative productivity in the development of west-European physics in the 18th and 19th centuries.' *Isis, 12*, 287–319.

Ranke, L. (1885). *Geschichte der romanischen und germanischen Völker von 1494 bis 1514*. Leipzig. First published 1824.

Reingold, N. (1981). 'Science, scientists, and historians of science.' *History of Science, 19*, 274–283.

Rescher, N. (1978). *Scientific Progress*. Oxford: Basil Blackwell.

Roller, H.D., ed. (1971). *Perspectives in the History of Science and Technology*. Norma (Oklahoma): University of Oklahoma Press.

Roll–Hansen, N. (1980). 'The controversy between biometricians and Mendelians: a test case for the sociology of scientific knowledge.' *Social Science Information, 19*, 501–517.

Rosen, N. (1960). 'Calvin's attitude towards Copernicus.' *Journal of the History of Ideas, 21*, 431–441.

Ross, S. (1962). 'Scientist: the story of a word.' *Annals of Science, 18*, 65–86.

Rothman, T. (1982). 'The short life of Évariste Galois.' *Scientific American*, April, 112–120.

Russell, C.A., ed. (1979). *Science and Religious Beliefs*. Sevenoaks (Kent): Open University.

Sachs, M. (1976). 'Maimonides, Spinoza and the field concept in physics.' *Journal of the History of Ideas, 37*, 125–131.

Sailor, D.B. (1964). 'Moses and atomism.' *Journal of the History of Ideas, 25*, 3–16. Reprinted in Russell (1979), pp. 5–19.

Sambursky, S. (1963). *The Physical World of the Greeks*. London: Routledge and Kegan Paul.

Sandler, I. (1979). 'Some reflections on the protean nature of the scientific precursor.' *History of Science, 17,* 170–190.

Sarton, G. (1927–1948). *An Introduction to the History of Science.* 3 vols., Baltimore: Williams and Wilkins.

Sarton, G. (1936). *The Study of the History of Science.* Cambridge (Mass.): Harvard University Press.

Sarton, G. (1948). *The Life of Science.* New York: Henry Schuman.

Sarton, G. (1952). *Horus. A Guide to the History of Science.* Waltham (Mass.): Chronica Botanica.

Schaff, A. (1977). *Historie og Sandhed.* Copenhagen: GMT (Danish translation of *Historia i Prawda,* Warsaw, 1970).

Schorlemmer, C. (1879). *The Rise and Development of Organic Chemistry.* Manchester.

Schrödinger, E. (1954). *Nature and the Greeks.* Cambridge: Cambridge University Press.

Seeger, R.J. (1965). 'Galileo, yesterday and today.' *American Journal of Physics, 32,* 680–698.

Segre, M. (1980). 'The role of experiment in Galileo's physics.' *Archive for History of the Exact Sciences, 23,* 227–252.

Shankland, R.S. (1963). 'Conversations with Albert Einstein.' *American Journal of Physics, 31,* 47–57.

Shapin, S. (1974). 'The audience for science in eighteenth century Edinburgh.' *History of Science, 12,* 95–121.

Shapin, S. (1975). 'Phrenological knowledge and the social structure of early nineteenth-century Edinburgh.' *Annals of Science, 32,* 219–243.

Shapin, S. (1982). 'History of science and its sociological reconstruction.' *History of Science, 20,* 157–211.

Shapin, S. (1984). 'Talking history: reflections on discourse analysis.' *Isis, 75,* 125–130.

Shapin, S. and Thackray, A. (1974). 'Prosopography as a research tool in history of science: the British scientific community 1700–1800.' *History of Science, 12,* 1–28.

Shea, W.R. (1972). *Galileo's Intellectual Revolution.* New York: Science History Publications.

Shea, W. (1977). 'Galileo and the justification of experiments.' In Butts and Hintikka (1977), pp. 81–92.

Shore, M.F. (1981). 'A psychoanalytic perspective.' *Journal of Interdisciplinary History, 12,* 89–113.

Simonton, D.K. (1976). 'The causal relation between war and scientific discovery.' *Journal of Cross-Cultural Psychology, 7,* 133–144.

Simonton, D.K. (1984). *Genius, Creativity and Leadership: Historiometric Inquiries.* Cambridge (Mass.): Harvard University Press.

Skinner, Q. (1969). 'Meaning and understanding in the history of ideas.' *History and Theory, 7,* 3–53.

Small, H.G. (1977). 'A co-citation model of a scientific speciality: a longitudinal study of collagen research.' *Social Studies of Science, 7*, 139–166.

Small, H.G., ed. (1981). *Physics Citation Index 1920–1929*. 2 vols., Philadelphia: Institute for Scientific Information.

Snow, C.P. (1966). *The Two Cultures and a Second Look*. Cambridge: Cambridge University Press.

Solla Price, D.J. de (1956). 'The exponential curve of science.' *Discovery, 17*, 240–243.

Solla Price, D.J. de (1963). *Little Science, Big Science*. New York: Columbia University Press.

Solla Price, D.J. de (1972). 'Science and technology: distinctions and interrelationships.' In Barnes (1972), pp. 166–180.

Solla Price, D.J. de (1974). *Science Since Babylon*. New York: Yale University Press.

Solla Price, D.J. de (1980). 'The analytical (quantitative) theory of science and its implications for the nature of scientific discovery.' In Grmek, Cohen and Cimino (1980), pp. 179–189.

Sorokin, P.A. (1937). *Social and Cultural Dynamics*. New York: American.

Spengler, O. (1926). *The Decline of the West*. London: Allen and Unwin.

Steffens, H. (1968). *Indledning til Philosophiske Forelæsninger*. Copenhagen: Gyldendal. First published Copenhagen, 1803.

Stephenson, R. (1982). 'The skies of Babylon.' *New Scientist*, 19 August, 478–481.

Stern, F., ed. (1956). *The Varieties of History*. New York: Meridian.

Stone, L. (1971). 'Prosopography.' *Dædalus*, winter, 46–79.

Streisler, E. *et al.*, eds. (1969). *Roads to Freedom: Essays in Honour of F.A.Hayek*. London: Routledge and Kegan Paul.

Stuewer, R.H., ed. (1970). *Historical and Philosophical Perspectives of Science*. Minneapolis: University of Minnesota Press.

Sudhoff, K., ed. (1910). *Klassiker der Medizin*. Leipzig.

Sullivan, D., White, D.H. and Barboni, E.J. (1977a). 'The state of a science: indicators in the specialty of weak interactions.' *Social Studies of Science, 7*, 167–200.

Sullivan, D., White, D.H. and Barboni, E.J. (1977b). 'Co-citation analyses of science: an evaluation.' *Social Studies of Science, 7*, 223–240.

Sullivan, D., White, D.H. and Barboni, E.J. (1979). 'The interdependence of theory and experiment in revolutionary science: the case of parity violation.' *Social Studies of Science, 9*, 303–327.

Suppe, F., ed. (1977). *The Structure of Scientific Theories*. Urbana (Illinois): University of Illinois Press.

Tannery, P. (1912–1950). *Mémoires scientifiques*. 17 vols., Paris: Gauthier–Villars.

Thackray, A. (1966). 'The origin of Dalton's chemical atomic theory: Daltonian doubts resolved.' *Isis, 57*, 35–55.

Thackray, A. (1972). *John Dalton. Critical Assessments of his Life and Science.* Cambridge (Mass.): Harvard University Press.

Thackray, A. (1974). 'Natural knowledge in a cultural context: the Manchester model.' *American Historical Review*, 79, 672-709.

Thackray, A. (1978). 'Measurement in the historiography of science.' In Elkana *et al.* (1978), pp. 11–30.

Thackray, A. (1980). 'History of science.' In Durbin (1980), pp. 3–69.

Thom, A. (1971). *Megalithic Lunar Observations.* Oxford: Oxford University Press.

Thomson, T. (1825). *An Attempt to Establish the First Principles of Chemistry by Experiment.* 2 vols., London.

Thomson, T. (1830–1831). *History of Chemistry.*, 2 vols., London.

Todhunter, I. (1861). *History of the Calculus of Variations During the Nineteenth Century.* Cambridge.

Todhunter, I. (1865). *History of the Mathematical Theory of Probability.* Cambridge.

Todhunter, I. (1873). *A History of the Mathematical Theories of Attraction and the Figure of the Earth.* Cambridge.

Truesdell, C. (1968). *Essays in the History of Mechanics.* Berlin: Springer–Verlag.

Truesdell, C. (1980). *The Tragicomical history of Thermodynamics 1822–1854.* Berlin: Springer–Verlag.

Vavilov, S.J. (1947). 'Newton and the atomic theory.' In *Newton Tercentenary Celebrations*, London: Royal Society of London.

Vickers, B., ed. (1984). *Occult and Scientific Mentalities in the Renaissance.* Cambridge: Cambridge University Press.

Vorzimmer, P.J. (1968). 'Darwin and Mendel: the historical connection.' *Isis, 59*, 77–82.

Walden, P. (1944). *Drei Jahrtausende Chemie.* Berlin: W.Limpert.

Watkins, J.W.N. (1953). 'Ideal types and historical explanation.' In Feigl and Brodbeck (1953). pp. 723–743.

Watkins, J.W.N. (1959). 'Historical explanation in the social sciences.' In Gardiner (1959), pp. 503–513.

Weiner, C., ed. (1977). *History of Twentieth Century Physics.* New York: Academic Press.

Wertheimer, M. (1959). *Productive Thinking.* New York: Harper.

Westfall, R.S. (1958). *Science and Religion in Seventeenth-Century England.* New Haven: Yale University Press.

Westfall, R.S. (1976). 'The changing world of the Newtonian Industry.' *Journal of the History of Ideas, 37*, 175–184.

Westfall, R.S. (1980). *Never at Rest. A Biography of Isaac Newton.* Cambridge: Cambridge University Press.

Weyer, J. (1972). 'Prinzipien und Methoden des Chemiehistorikers.' *Chemie in unserer Zeit, 6*, 185–190.

Weyer, J. (1974). *Chemiegeschichtsschreibung von Wiegleb (1790) bis Partington (1970)*. Hildesheim: Gerstenberg.

Whewell, W. (1837). *History of the Inductive Sciences*. 3 vols., London. Reprinted London: Cass, 1967.

Whewell, W. (1840). *The Philosophy of the Inductive Sciences, Founded upon their History*. 2 vols., London.

Whewell, W. (1867). 'On the influence of the history of science upon intellectual education.' In Youmans (1867), pp. 163–189.

Whitaker, M. (1979). 'History or quasi-history in physics education.' *Physics Education, 14*, 108–112.

White, A.D. (1955). *A History of the Warfare of Science with Theology in Christendom*. London: Arco Publishers. First published New York, 1896.

Wightman, W. (1951). *The Growth of Scientific Ideas*. Edinburgh: Oliver and Boyd.

Wohlwill, E. (1909). *Galilei und sein Kampf für die Copernicanische Lehre*. 2 vols., Hamburg.

Wolff, M. (1978). *Geschichte der Impetustheorie*. Frankfurt: Suhrkamp.

Wood, P. (1983). 'Philosophy of science in relation to history of science.' In Corsi and Weindling (1983), pp. 116–135.

Woodall, A.J. (1967). 'Science history – the place of the history of science in science teaching.' *Physics Education, 2*, 297–305.

Woolgar, S.W. (1976). 'Writing an intellectual history of scientific development: the use of discovery accounts.' *Social Studies of Science, 6*, 395–422.

Worrall, J. (1976). 'Thomas Young and the "refutation" of Newtonian optics: a case-study in the interaction of philosophy of science and history of science.' In Howson (1976), pp. 181–210.

Youmans, E.L., ed. (1867). *Modern Culture*. London.

Young, T. (1802). 'On the theory of light and colour.' *Philosophical Transactions of the Royal Society of London, 92*, 12–24.

Yuasa, M. (1962). 'Center of scientific activity: its shift from the 16th to the 20th century.' *Japanese Studies in the History of Science, 1*, 57–75.

Zirkle, C. (1964). 'Some oddities in the delayed discovery of Mendelism.' *Journal of Heredity, 55*, 65–72.

Index

....

Acton, Lord, 41
Agassi, J., 168
Althusser, L., 108
Ampère, A.M., 154f
anachronisms, 75, 79, 89
anticipation, 100f
archaeometry, 166
Aristotle, 1, 96
authenticity, 129, 162
Avogadro, A., 80

Bachelard, G., 92f
Bacon, F., 5
Bailly, J.-S., 3, 6
Beard, C., 44–6
Becker, C.L., 46
Belloni, L., 159
Bernal, J.D., 38, 71
Beyerchen, A., 24
biographies, 168–73
Biot, J.-B., 100
Blake, C., 56
Bloch, M., 54, 131
Blondlot, R., 67
Boas Hall, M., 27
Bohr, N., 192
Boscovich, R.J., 102
Boyle, R., 3, 28
Brannigan, A., 68
Buchdal, G., 91

Calvin, J., 135–6
Candolle, A. de, 175
Caneva, K.L., 124
Canguilhem, G., 73, 92
Cannizaro, S., 80
Carr, E.H., 43–4, 47
Chryssipos, 86
citation analysis, 191–5f
Clagett, M., 16
Clark, J.T., 37
co-citation analysis, 195

Cohen, I.B., 27, 38, 96
coherence, mythology of, 97, 165
Cole, F.J., 182
Collingwood, R.G., 30, 48–50, 90
Comte, A., 11–13, 17
Conant, J., 34
Copernicus, N., 82, 136, 156
counterfactuals, 70
Croce, B., 47f
Crombie, A.C., 16, 76
Crosland, M., 138
Crowther, J.G., 135
Curie, E., 169

Dahl, 0., 123
Dalton, J., 138–43
Danneman, F., 17
Danto, A.C., 46
Darmstaedter, L., 17, 175
Darwin, C., 8, 52, 194
De Groot, H., 148
Delbrück, M., 176
Dewey, J., 47
Dijksterhuis, E.J., 30, 145, 148, 161
DN model, 62,–5
Dobbs, B., 27
Dobrov, G., 187
Drake, S., 144–8
Dray, W.H., 65
Du Bois-Reymond, E., 14
Duhem, P., 16, 36, 76, 103, 110, 155
Durant, W., 135

Eames, N.B., 182
Edge, D., 196
Einstein, A., 50, 73, 115, 151–4
Elie de Beaumont, L., 79
Elkana, Y., 40
Elliott, C.A., 166
emergence technique, 103
Engels, F., 13

Index 235

Tannery, P., 15, 17
Thackray, A., 143, 158, 178f, 184
thematic analysis, 85
Thomson, T., 139f
Todhunter, I., 8
translation, 134f
Truesdell, C., 32, 89, 146, 148
truth, historical, 58
two cultures, 37

unit-ideas, 83f
Vico, G., 5
Viviani, V., 147
Vrain-Lucas, 129

Watkins, J.W., 68–9
Werner, A., 45
Westfall, R., 27, 170
Whewell, W., 5, 8, 9, 25, 111
White, A., 135
Whiteside, D.T., 27
Wöhler, F., 164
Wohlwill, E., 16
Wolff, M., 88
Woolgar, S.W., 156–8
working history, 112

Young, T., 113